HANDS-ON HOME LANDSCAPING
from THE PAVING GUY

HANDS-ON HOME LANDSCAPING
from THE PAVING GUY

NIGEL RUCK

PUBLISHED BY JAMIE DURIE PUBLISHING

JPD MEDIA PTY LTD
ABN 83 098 894 761
35 Albany Street
Crows Nest NSW 2065
PHONE: + 61 2 9026 7444
FAX: + 61 2 9026 7475

FOUNDER AND EDITORIAL DIRECTOR: Jamie Durie
GROUP CREATIVE DIRECTOR: Nadine Bush

PUBLISHER: Nicola Hartley
MANAGING EDITOR: Bettina Hodgson
PUBLISHING SERVICES MANAGER: Belinda Smithyman
CREATIVE DIRECTOR: Amanda Emmerson
DESIGN CONCEPT AND ART DIRECTION: Amanda Emmerson
DESIGN: Criena Court, Sonia McAllan
EDITOR: Edwin Barnard
CONTRIBUTING EDITOR: John Monty
PROOFREADER: Catherine J Page
PHOTOGRAPHY: Jason Busch
STYLING: Marcus Hay
ILLUSTRATIONS: Giselle Barron, Grace Mansour

DISTRIBUTED BY: HarperCollins*Publishers* Australia

© 2006 TEXT: Nigel Ruck
© 2006 CONCEPT, DESIGN AND PHOTOGRAPHY: Jamie Durie Publishing

All rights reserved. No part of this book may be reproduced or transmitted in any form or by any means, electronic or mechanical, including photocopying, recording, or by any information storage and retrieval system, without permission in writing from the publisher.

DISCLAIMER: Any person who uses the information contained in this book to carry out any design or landscaping projects does so at their own risk and the Publisher and the Author, to the extent permitted by law, do not accept any responsibility or liability for any loss or damage which any person may suffer arising from use of the information contained in this book including any damage to property or personal injury. While the Publisher and the Author have used their best efforts in preparing this book, they make no representations or warranties with respect to the accuracy or completeness of the contents of this book. The advice and projects may not be suitable for your individual requirements and you should not embark on an activity that might reasonably be outside your skills or ability. You should consult with a professional where appropriate. Before undertaking any project you are directed to the "Safety First" section of the book which identifies certain risks, clothing and equipment.

National Library of Australia Cataloguing-in-Publication data:
Ruck, Nigel
Hands-on Home Landscaping from The Paving Guy

Includes index.
ISBN 0 9757361 0 8
1. Landscape design – Amateurs' manuals. 2. Garden structures – Design and construction – Amateurs' manuals.
I. Title.
712.6

Set in Gotham on InDesign
Printed in Singapore by Tien Wah Press
First printed in 2006

10 9 8 7 6 5 4 3 2 1

* There are references to *Backyard Blitz* within the text of this book. *Backyard Blitz* is a registered trade mark of CTC Productions Pty Limited. This publication is in no way affiliated with CTC Productions Pty Limited, the producers of *Backyard Blitz*.

ADDITIONAL PHOTOGRAPHY
PAGES 18–23, 37–41, 44–46, 59 *(top right, bottom left)*, **60** *(top left)*, **95** *(top left and bottom right)*, **111** *(bottom right)*: Leigh Clapp
PAGES 12–17, 42 *(bottom left)*, **60** *(top right)*, **111** *(top right)*: Jim Fogarty
PAGE 34 *(top right and bottom right)*: Secret Gardens of Sydney
PAGE 42 *(top right)*: Danny Kildare
PAGE 59 *(bottom right)*: Harriette Rowe

GARDEN DESIGN
PAGES 12–17, 42 *(bottom left)*, **60** *(top right)*, **111** *(top right)*: Jim Fogarty
PAGES 18–23: David Baptiste
PAGE 38 *(top right)*: Andrew Prowse
PAGE 34 *(top right and bottom right)*, **42** *(top right)*: Secret Gardens of Sydney, www.secretgardens.com.au
PAGE 46 *(top right)*: Taylor Cullity Lethlean Landscape Architects
PAGE 46 *(bottom left)*: Suzan Slater
PAGE 59 *(bottom right)*: PATIO Landscape Architecture & Design
PAGE 59 *(bottom left)*, **60** *(top left)*: Peter Fudge

THE PUBLISHERS WOULD LIKE TO THANK
For use of garden spaces: Tony and Janet Camilleri, Kemble and Katrina Cowan, Andrew Barlow and Kristel Shaw, Sophie Bartho and Rob Wilcher, Rob and Judy Connor, John and Elspeth Thixton, Jason and Isobel Gram, James Economides, Kevin Bishop and Daniel Vucetich, Phil and Jenny Currie, Michael and Suzie Blumentals, Di Englander, Brett and Jenny Parsons, John and Maggie Cechin, Capse Lodge and Michael Cooke.

Our wonderful talent: Lily Ruck, Belinda Smithyman, Ashley Gualter and Milli.

For illustration and image assistance: Jennifer Anderson, Simon Howard, Tracy Loughlin, Jim Fogarty, Melanie Burgess, Rochelle Abood and Matt Cantwell.

With very special thanks for support to:

 Makita Power Tools, www.makita.com.au

 KISSS Sub-Surface Irrigation Systems, www.KISSS.net.au

 Riverstone Pavers, www.riverstonepavers.com.au

 Patio by Jamie Durie products, available in all Kmart stores, www.kmart.com.au

PROPS
Positively Curlewis Street: **T** (02) 9365 6000
Country Road: www.countryroad.com.au **T** (02) 9369 3420, 1800 801 911
IKEA: www.ikea.com.au **T** (02) 9313 6400

FOREWORD

FOR A FEW YEARS NOW I HAVE PUBLISHED BOOKS TO INSPIRE people in some way shape or form. I take great joy in photographing the gardens we create — making sure that we shoot them at just the right time of year when the plants are in flower and the leaves are looking lush, and at just the right time of day when the light feels soft and the gardens are at their most inviting. We do this with the purpose of connecting people with nature in the most inspirational way possible, because every person we entice into the wonderful world of garden creation is another plant in the ground for our planet and another person who has learned just that little bit more about nature.

Now while all of this sounds great, it can mean nothing without sufficient follow-through and the knowledge and tools to make a garden happen. I am delighted that finally we've come up with a practical, hands-on book that teaches you step by step how to put all of this garden inspiration into action. Nigel, apart from being a great mate of mine, has been my colleague for almost 7 years. We have shared many stories and even more laughs in the backyard and created some stunning gardens along the way for some incredible people, but none of this would have happened without Nigel's fantastic knowledge and construction experience. This book helps you create simple, affordable, quality-built gardens that will only get better with time. It's for the seasoned gardener, the weekend warrior, the novice and even your mum or dad!

Let us know how you go...happy landscaping.

Jamie Durie

CONTENTS

WHO IS THE PAVING GUY? Meet landscape gardener and author Nigel Ruck. **1**

CREATING A GARDEN How to choose the best solution for your site. **5**

LET'S MAKE A PLAN Translating your ideas into a workable plan. **31**

DOING THE HARD GRAFT Practical skills that will turn your plan into reality. **69**

CAN'T DO WITHOUT IT Tools and a useful reference for the home landscaper. **121**

Index **134**

Acknowledgements **136**

WHO IS THE PAVING GUY?

1

NIGEL RUCK HAS MADE A NAME FOR HIMSELF ON THE POPULAR AND LONG RUNNING TELEVISION PROGRAM *BACKYARD BLITZ** AS 'THE PAVING GUY'. HIS CONSIDERABLE EXPERTISE EXTENDS BEYOND JUST PAVING, HOWEVER, TO COVER ALL ASPECTS OF THE ART AND PRACTICE OF HOME LANDSCAPING.

INTRODUCTION

I FIRST CAME TO AUSTRALIA IN 1988, AS PART OF A GLOBAL working holiday that went on for about seven years — a touch longer than the 'year off' I vaguely had in mind when I left England. My first landscaping job was in Whistler, Canada, but at the time I was really there for the skiing, and the parties of course. It was only after I settled in Australia that I decided I would enjoy a career in landscaping, and for the next three years I delivered (and ate) a truckload of pizzas while studying at Ryde TAFE. In 1996 I started out on my own as a landscaping contractor, and by 1999 I had built up a thriving little business.

But I've never been one to stand still for long. I'm always on the lookout for new opportunities, and when a television producer mate of mine rang me one night, here was one opportunity that seemed just too good to miss. Before I knew it I was in front of the camera doing a screen test for a new make-over show, *Backyard Blitz**, and a couple of weeks later they rang to say the job was mine.

That was seven years ago. The show is still going strong, and now I'm writing my first book, something I often thought about while I was travelling. Well here it is — not the confessions of a party animal, but a guide to home landscaping. Of course, there's no way of including everything there is to know about landscaping in 144 pages. What I've tried to do instead is probably more useful, and that is to lead you through the whole process — from deciding on a style to putting the finishing touches to the garden of your dreams.

One final word of advice: landscaping is fun, so just make sure you enjoy yourself!

CREATING A GARDEN

EVERY GARDEN IS DIFFERENT: A CREATION REFLECTING THE DEMANDS OF THE SITE, THE TASTES AND NEEDS OF ITS OWNER AND THE SKILLS AND IMAGINATION OF THOSE WHO MADE IT. HERE WE EXPLORE HOW THOSE FACTORS COMBINED IN THE CREATION OF FOUR INTERESTING GARDENS.

2

COURTYARD CHIC

ENTERTAINING, LOW MAINTENANCE AND SPACE for children were high priorities for the young couple who asked me to take a look at the garden behind their recently purchased suburban cottage. Situated not far from Sydney's popular Manly beach, the street where the house stood had a pleasant, relaxed holiday atmosphere, with couples and families strolling past on their way down for a swim.

There was certainly plenty of room for improvement. The site was on the small side and the street behind, where you entered the garage, was quite a bit higher than the house itself. This meant that the first thing that grabbed your attention when you walked into the garden through the French doors was a blank concrete-block wall and an ugly staircase at the end of the yard. That, and a washing line firmly cemented into the middle of the garden, were among the first things I mentally noted down for attention. Apart from that, there was some basic landscaping work that had been done a few years ago, but it was starting to look very tired and was well overdue for replacement.

On the plus side, the site had a number of good points in its favour. Most of the garden received plenty of sun throughout the year, a fact confirmed by a very healthy looking frangipani that flourished beside one fence. Privacy was not a big

Only a couple of seasons of growth are needed for the garden to take on its finished look. The outdoor shower *(near left)* is for the use of bathers returning from the beach.

(Clockwise from below left) New seating in the cabana provides a shady summer retreat. The mature frangipani was one of the garden's most important existing assets. A single socket now serves both an umbrella and the removable clothes hoist. Shoppers carrying bulky packages need a generous turning space at the base of the garage steps. Work gets underway in the garden as it was originally. The view back towards the house from the top of the garage steps.

issue, especially as trees in neighbouring gardens provided some screening. The garden was also shielded to some extent from prevailing winds — an important point when it came to choosing plants, because there was an ocean beach only a couple of blocks away and that meant there could be some salt in the air, especially during severe storms.

The clients had given a lot of careful thought to what they wanted — which was essentially something with clean, uncluttered modern lines. Since they were going to use the garden mainly for entertaining, they were keen to retain an existing cabana and its barbecue. Paving rather than turf was the preferred option for a ground finish. Although not as soft as turf, it would still work as an outdoor play space for the owners' baby boy when he was a bit older.

Our first job was to strip everything out of the garden except the cabana. The existing garden beds were demolished, as were the ugly concrete steps leading down from the garage. Since the timber boundary fences were sound, these were retained, to be repainted later.

The general plan was to replace the planters down the sides of the garden and then install another planter about two-thirds of the way down towards the cabana. This new central planter would provide a focal point for the

PG *recommends*
DON'T FORGET THE LIGHTING

I'm a great fan of outdoor lighting, which can completely change the feel of a garden. It was particularly important here, since the area was to be used for both day- and night-time entertaining. The main features of the low-voltage system we installed are strategically placed eyelid lights to help users navigate paths and steps, and carefully angled spots to highlight the yuccas.

CREATING A GARDEN

garden — particularly at night when the plants in it would be lit by concealed spots. It would also help to split the garden into two spaces, with the aim of creating a slight air of mystery. A garden is more interesting if you can't see everything in it at a single glance.

Next we poured concrete footings for the planters, and installed subsurface drainage inside the planters and under the area to be paved. We wanted to keep the main area of paving as clean and neat as possible, and this effect would have been spoilt if water was constantly weeping from holes at the base of the planters. We also installed wiring for the planned low-voltage lighting system.

With all the services and preparation work complete, we could now build and render the planters, install the eyelid lights, bring in the soil and position the plants we had chosen. With the plants, we were aiming for strong contrasts in shape and texture, with dense, soft, weeping foliage on either side of the garden and three striking yuccas in the central planter, where their strong architectural lines immediately attracted the eye.

Finally, timber garage steps were built, the back wall rendered, new seating installed in the cabana and a final coat of paint applied where necessary. The paving was left till last so it would not get stained by soil during planting. For this we chose 400 x 400mm pavers, laid on the diagonal to create interest.

what we planted...

- **RIGHT-HAND BED:** *Howea forsteriana, Murraya paniculata, Cycas revoluta, Liriope muscari* 'Evergreen Giant'.
- **LEFT-HAND BED:** existing frangipani (*Plumeria* spp.), *Gardenia augusta* 'Florida', *Liriope muscari, Syzygium luehmannii, Archontophoenix* spp., *Howea forsteriana, Ophiopogon japonicus*.
- **CENTRAL PLANTER:** *Yucca elephantipes, Ophiopogon japonicus*.
- **SQUARE POTS BEHIND CENTRAL PLANTER:** *Cordyline rubra, Trachelospermum jasminoides*.
- **SQUARE POTS BESIDE CABANA AND GARAGE STEPS:** *Agave attenuata*.
- **REAR CABANA SEAT:** *Liriope muscari*.

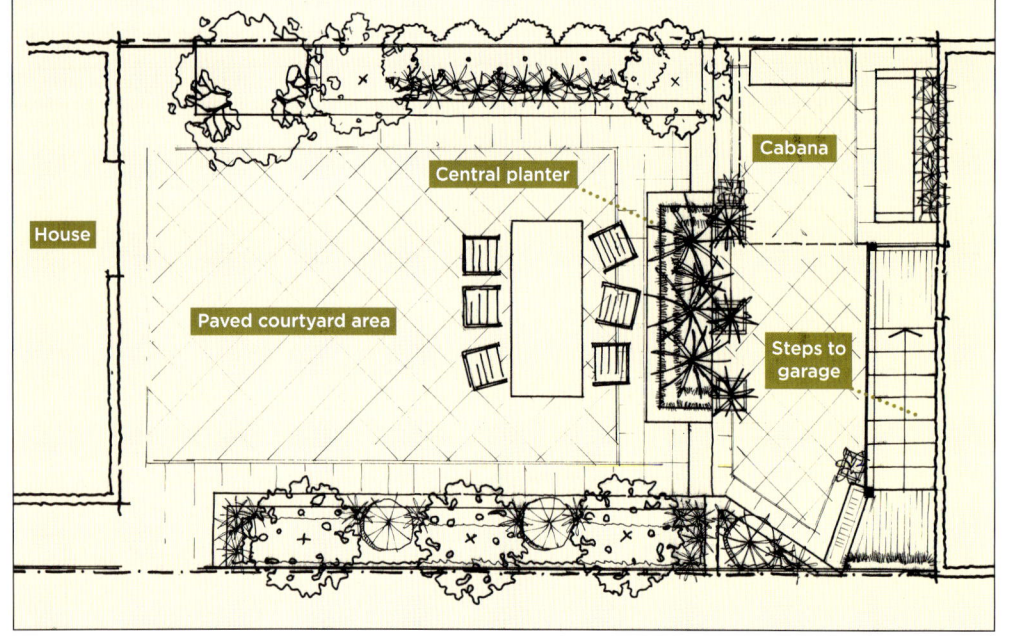

The striking shapes and patterns created by the trunks and foliage of the yuccas in the central planter are accentuated at night by light from three concealed spots.

A sketch plan of the garden showing the positions of the main landscaping elements.

AN ISLAND OF CALM IN AN OCEAN OF WIND

A HIGHLY ORIGINAL HOUSE AND A CHALLENGING environment were the two most important factors that played a part in influencing the design of this unusual garden. Arriving at the right solution was vital in ensuring the overall success of the house as a place where the family that commissioned it could live in comfort.

The house, designed by an award-winning firm of Australian architects, was built in a predominantly rural area on Victoria's Mornington Peninsula, not far from the ocean. The design makes extensive use of recycled materials, such as old wharf timbers, which are combined with steel and glass to give the house a strongly contemporary feel. Lime green was chosen by the architects as a feature colour on parts of the exterior of the house, so any landscaping elements had to be chosen with care to avoid colour clashes.

What makes the environment particularly challenging is the seemingly ever-present wind, which blows mainly from the south. The area's annual average wind speed approaches 20km/h, and monthly averages remain fairly constant throughout the year.

A feature of this design is the way in which new elements, such as the jarrah benches *(here and far left)* and the exposed aggregate paving *(here and near left)* harmonise with existing materials, such as the recycled timber used for the house.

(Clockwise from below left) The use of natural materials and finishes helps to impart a rustic feel to the garden. Lime green feature walls on the house strongly influenced the choice of materials and plants. Suspended screens help to further reduce wind speed under the house. Walls and other structures create sheltered corners where plants can thrive. The area under the house before its transformation. A characteristic of rammed earth is its ability to change colour dramatically in different light conditions. A view past the house showing the character of the surrounding landscape.

The brief from the owners was straightforward. They had recently moved into the house and wanted family space outdoors, an area for entertaining friends in summer and, above all, some protection from the area's strong winds.

Wind was certainly always going to be the major challenge for any landscape designer who took on this job. It seemed to rule out an extensive garden reaching out away from the house because of the obvious difficulty and cost of providing protection for plants and users. Apart from the wind factor, however, it was also important to come up with a design that harmonised with the building's rural surroundings. Extensive plantings of trees and shrubs would have created an 'island' of vegetation that would have seemed out of character with the natural environment. Also, such a solution would have tended to detract from the bold statement that the building makes in its setting.

Since an extensive garden seemed to be out of the question, attention turned to a vacant area under the house instead. Here an existing concrete wall gives some protection, but only from wind from certain directions. To increase the protection, some free-standing rammed earth walls were built to create wind breaks, but without totally enclosing the space.

Rammed earth is a very old technology for which you can use the local soil, either with or without

PG *recommends*
THINKING THINGS THROUGH
This garden is a great example of the need to come up with a solution that addresses the various challenges a site presents. A conventional garden with lawns and flower beds would not only have looked out of place here, but it would have been entirely impractical. When something so contemporary imposes on the environment, balance it out by using local materials.

CREATING A GARDEN

additives, depending on its consistency and whether it contains a lot of clay or sand. The soil is shovelled into stout timber forms in layers and rammed down as hard as possible, using either hand rammers or machines. As each layer is compressed, the forms are moved up until the wall reaches the desired height. Provided it is well protected, rammed earth is astonishingly durable, and walls several hundred years old are known to be still standing in parts of Europe.

Further protection against wind for the family space that was being created on the north side of the house was provided by screens made from horizontal timber slats and perforated sheet metal. These were suspended at strategic points in breezeways to further reduce wind velocity to comfortable levels.

Once the space was adequately protected, work could begin on other elements of the design. Planters were created at the base of walls and curved jarrah benches constructed to provide seating. Ground finishes combined natural stone crazy paving and concrete with an exposed aggregate finish using local crushed granite. An existing concrete wall was painted to match the colour and texture of the new rammed earth walls.

With building complete, all that remained to do was the planting. Here the plan was for a very simple scheme using only a small number of hardy species designed to create a suitably rustic feel.

what we planted...

Plantings in this garden were kept deliberately sparse to reflect the nature of the surrounding landscape and to complement the spare, almost austere character of the building itself and its new garden. Most species were chosen for their ability to withstand harsh conditions, and as a result the garden requires very little maintenance and only occasional watering.

- **BASE OF LONG WALL:** *Arthropodium cirrhatum*.
- **BASE OF RAMMED EARTH WALLS:** *Pennisetum* spp. and *Carex* spp. (the latter recognisable by their distinctive coppery-brown tones).
- **LARGE BED:** *Nepeta* spp.

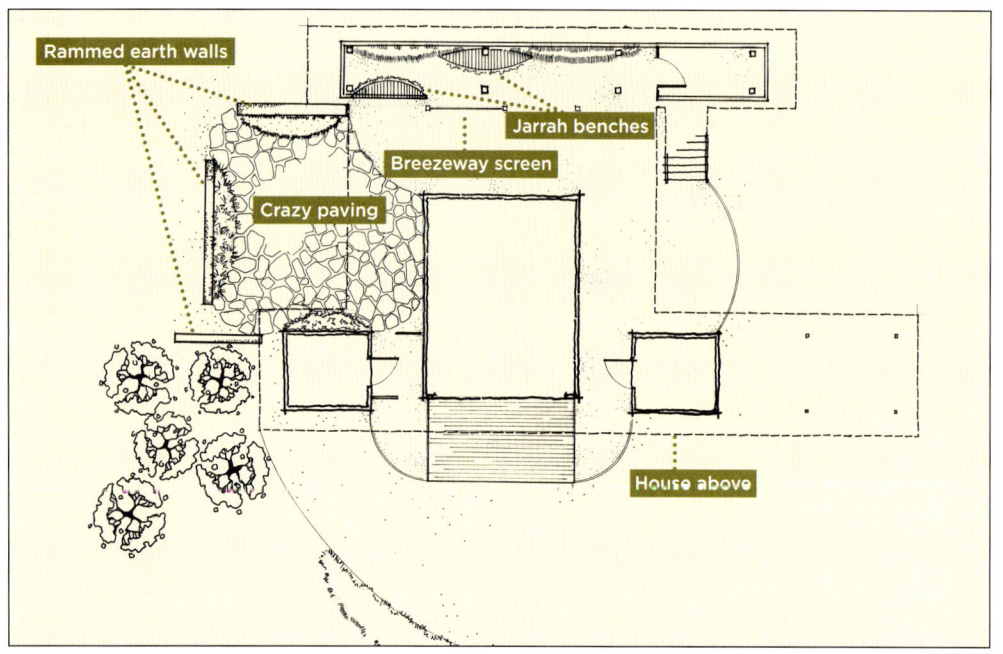

By using walls, screens and other structures, the designer has created a variety of areas around the garden where users can sit in comfort whatever the conditions.

A sketch plan of the garden showing the positions of the main landscaping elements.

CLEVER SOLUTIONS IN AN URBAN OASIS

PLANNING FOR ENERGY EFFICIENCY WAS ALWAYS a priority for the owners of this garden, right from the time when they bought the block of land for their dream home in a suburb of Adelaide. The land wasn't very big — it had once been a tennis court — so careful planning was going to be essential if the best use was to be made of every square metre. There was also the climate to consider. Although Adelaide enjoys generally mild weather for most of the year, hot, dry spells are often a feature of summer, when temperatures top 40°C or more. Winters can also be decidedly chilly.

The briefs for both architect and garden designer contained essentially the same message — create an integrated solution for the house and the garden that will reduce the need for heating and cooling to a minimum. For the house, this aim was realised in features such as high thermal mass walls to retain heat in winter, careful positioning of windows, an open-plan design to encourage air movement,

Deciduous trees between the pond and the house provide shade for the house and patio in summer, but let sunlight through in winter. The combination of natural *(far left)* and industrial *(near left)* materials gives the garden a contemporary feel.

(Clockwise from left) Star jasmine provides both a colourful display on a pergola and a dense screen for a boundary *(above)*. Water lilies and Japanese iris flourish in pots plunge-planted in the large pond. Rainwater runoff from the house is collected in a 4000-litre tank at the side of the house, with any overflow piped underground to the ponds.

a steep roof to facilitate rainwater collection and wide eaves to control sunlight — keeping it out in summer, when the sun is high in the sky, but allowing some through in winter, when the sun is lower.

In the garden, centrepieces of the design are two ponds fed with rainwater gathered from the roofs of the house and garage. They are positioned at the front of the block to take advantage of regular afternoon breezes in summer to waft cool, moist air through the house. Extensive plantings around the ponds and the rest of the house provide shade, privacy and some protection from street noise.

Since the ponds are the largest structural components of the landscape, with other elements being built around them, they were first to be installed. A backhoe was used to excavate shallow depressions, which were surfaced with a layer of sand to provide a soft finish and then waterproofed with custom-made butyl rubber liners. Pipes were installed to carry overflow from rainwater tanks at the back of the house, and then excavated soil was backfilled around the perimeter of the ponds and over the edges of the liners to create a waterside planting area. Pebbles and feature rocks were arranged to mimic as closely as possible what you would expect to find in a natural setting. The rest of the soil that had been excavated from the pond areas

PG *recommends*

PLAN FOR THE FUTURE

Energy-efficient buildings and gardens make more and more sense in Australia, especially in view of current concerns about urban pollution and the impending water crisis in many of our larger cities. This house is a fine example of what can be done to conserve precious resources in an urban environment, while at the same time creating a living space that is both practical and attractive.

CREATING A GARDEN

was mounded up to create some interesting variation in height on what is otherwise a very flat block.

With the major excavations out of the way, work could now begin on the remaining structural elements of the landscape design — a small paved entertaining and seating area beside the large pond, and paths where necessary around the perimeter of the building — as well as the extensive planting scheme.

A great deal of careful thought went into selecting plants and siting them where they would make an effective contribution to the overall plan. Deciduous trees — honey locusts, Manchurian and Bradford pear trees and quince — were chosen for areas around the house that need shade in summer and sun in winter. Evergreens such as Bull Bay magnolia, viburnums and evergreen ash provide privacy year-round and help to block out wind and noise. A dense understorey of shrubs and ground plants create a cool, moist, green environment beneath the trees, where the humidity remains well above that of the surrounding area, thanks to the presence of the ponds.

The finished garden has more than met the owner's expectations. Not only is it a pleasant place to be — especially on hot summer days when the temperature beneath the trees can be 10° lower than it is outside on the street — but it has also helped to make the house itself extremely livable in all seasons of the year.

what we planted...

- **ENTRANCE TO THE BLOCK:** *Pyrus calleryana* (Bradford pear).
- **ALONG FRONT OF BLOCK:** *Fraxinus griffithii, Viburnum tinus, Viburnum suspensum.*
- **AROUND POND:** *Gleditsia triacanthos* (honey locust), *Cydonia oblonga* (quince).
- **BORDERING GARAGE:** *Malus* spp. (espaliered).
- **EASTERN BOUNDARY:** *Magnolia grandiflora.*
- **BEHIND HOUSE:** *Pyrus ussuriensis* (Manchurian pear, pollarded).
- **IN POND:** *Nymphaea* spp., *Iris ensata.*
- **ARBOUR APPROACHING FRONT DOOR:** *Trachelospermum jasminoides.*

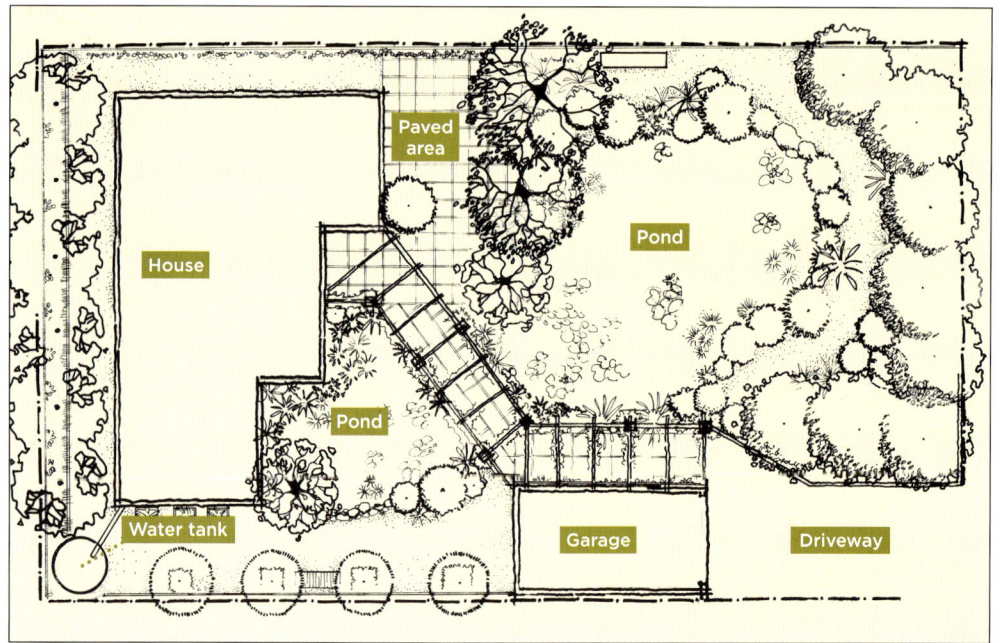

Manchurian pears along the back of the house are pollarded to provide shade for upstairs rooms in summer, but still let light through in winter when branches are bare.

A sketch plan of the garden showing the positions of the main landscaping elements.

MAKING THE MOST OF A SMALL SPACE

SUBURBAN GARDENS ALWAYS PRESENT a challenge for the landscaper. The average 'quarter-acre block' doesn't have a lot of room for a modern home as well as an interesting garden, particularly one that provides for the needs of a growing family whose members might also include pets. There are also neighbouring gardens and houses to consider when drawing up a plan.

When the owners purchased this block of land in an old established area, complete with its existing 1940s house, they already had a fairly clear idea of what they wanted to do. First they planned to demolish the existing building and replace it with a more contemporary home, then they wanted the garden completely remodelled in a semi-formal style, but with some Asian touches. The house design incorporated an outdoor room as part of its structure, so the landscape design would need to provide a transition between that and the garden proper.

The site presents few problems from a landscaping perspective. It has a gradual slope from front to back, which is an advantage, both from the point of view of drainage and also because it gives the site a more interesting character.

Gravel beds and timber pontoons create an effective link between the outdoor entertaining area and the lawn. Overall, the garden makes effective use of simple materials and plantings *(above and left)* to achieve the desired effect.

(Clockwise from below left) Careful pruning over time will fashion the box hedge into a third, living tier for the retaining wall. The two neatly trimmed *Ficus* in pots cast constantly changing shadows onto the wall of the house. The overall design of the garden echoes the symmetry and regularity of the house. Dark timber and pale gravel contrast strikingly in colour and texture. Bare earth and builder's rubble greet the landscapers on day one. A striking Balinese head acts as a focal point beside the entry steps. Water — here welling up from the interior of a large glazed pot — helps to create an air of tranquillity in the garden.

The land receives plenty of sun, and neighbours are far enough away for privacy not to be a major concern. The soil does contain a fair amount of clay, which means that any footings need to be substantial.

With the new house set well back from and below the level of the street, there was plenty of space to create an interesting entryway, and it was here that some of the garden's major structural elements were installed. The plan was for steps leading down from the road to an entry path, with curved, flanking retaining walls and a driveway surfaced with bluestone cobbles off to one side.

All of these elements called for substantial footings which, in the case of the retaining walls, had to be 1m wide. Since such wide footings left insufficient soil close to the wall for plants to thrive, a boardwalk was installed along its base, over the footings — a neat solution to a problem that could not be avoided. The boardwalk complements pontoons on either side of the steps, where they serve as platforms for a pair of cycads.

An extensive drainage system had to be installed behind the retaining walls, at the bottom of the drive and in front of the house slab, as well as a reservoir and pipework for a water feature and wiring for outdoor lighting. With the footings and other sub-surface work complete work could then begin on building the walls, steps,

PG *recommends*
ONE EYE ON THE BIG PICTURE
It's always important when planning a garden to make it look as if it 'belongs' in its environment. This involves choosing materials with colours and textures that complement existing structures, as well as plants that reflect the overall character of the landscape you are trying to create. When introducing themes — such as the Asian theme here — a few light touches are all that is needed.

CREATING A GARDEN

driveway and paths. Care had been taken to choose materials with textures and colours that harmonised with the house, such as the bricks and render on the curved retaining walls, which match those on the house, and the bluestone cobbles on the driveway that are the same colour as the roof tiles.

There was less structural work to do at the sides and back of the house, where the main focus was on providing a transition from the outdoor room to the garden. Here, as in the front garden, gravel-filled beds were constructed, with timber pontoons and lava stone steppers providing access. This area also includes a small water feature, made up of a decorative pot placed in a bed of small rocks. Water pumped into the base of the pot overflows down its sides and into a concealed reservoir, from where it is recycled. Small decorative touches, such as a sculpted head in a bed of slate chips and a few small stone statues help to reinforce the Asian/Japanese feel created by the neat beds of white pebbles.

The planting scheme for the garden is restrained, with no large trees or shrubs, apart from hedges along side and back boundaries. Box hedging edges the entry path and the top of the main curved retaining wall, while most of the other substantial plants — such as cycads flanking the entry steps — are in pots. Apart from the lawn and hedges, the resulting garden is low maintenance.

what we planted...

- **EDGING FOR ENTRY PATH AND CIRCULAR RETAINING WALL:** *Buxus* spp.
- **ENTRY STEP POTS:** *Cycas revoluta*.
- **RIGHT-HAND SIDE OF ENTRY STEPS:** *Liriope muscari*.
- **BED WITH SLATE CHIPS:** *Phormium tenax* 'Elfin'.
- **BED BEHIND STEPPED SIDE WALL:** *Syzygium luehmannii* 'Cascade'.
- **GRAVEL BED WITH TIMBER PONTOONS:** *Ophiopogon japonicus, Ophiopogon planiscapus* 'Nigrescens', *Buxus microphylla* var. *japonica, Dracaena marginata, Xanthorrhoea* spp.
- **BACK HEDGE:** *Syzygium australe* 'Hunchy'.
- **LAWN:** Semi-dwarf buffalo (ST26).

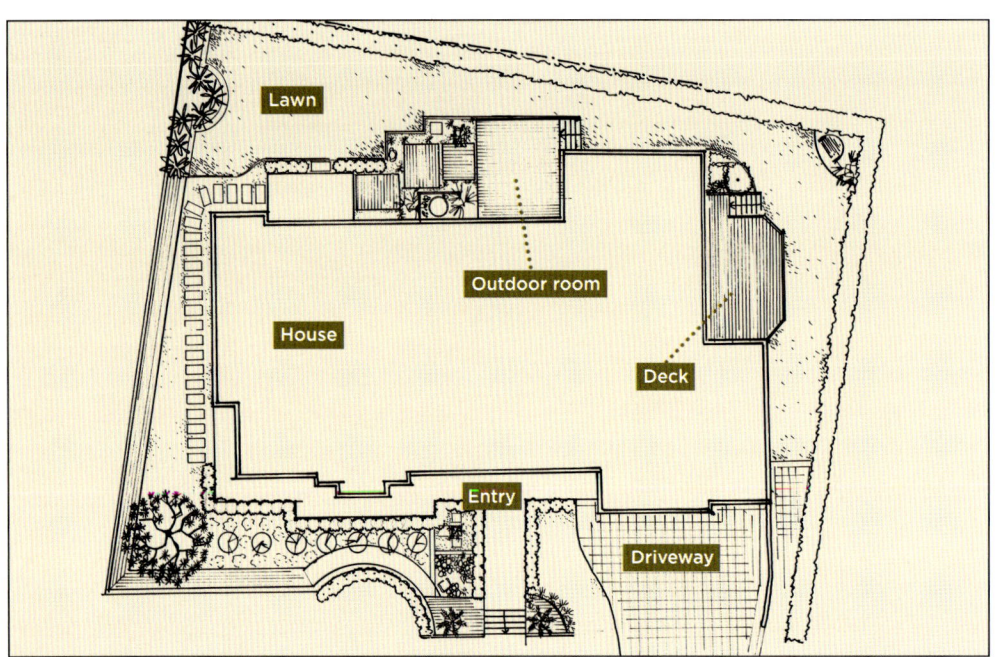

Over time, the box hedging along the side of the house will be clipped to form a solid mass about 400mm high.

A sketch plan of the garden showing the positions of the main landscaping elements.

LET'S MAKE A PLAN | 3

A CAREFULLY THOUGHT-OUT AND WORKABLE PLAN IS HALF THE BATTLE WHEN IT COMES TO CREATING THE GARDEN OF YOUR DREAMS. THERE ARE MANY FACTORS THAT HAVE TO BE CONSIDERED, AND IN THIS CHAPTER WE EXAMINE EACH OF THESE VARIOUS AND IMPORTANT STAGES IN DETAIL.

LET'S MAKE A PLAN

DECIDING ON WHAT YOU WANT

Where to look for sources of inspiration

Faced with the task of landscaping their garden, most home owners have trouble knowing where to start. This is hardly surprising, since garden design is not a job that most of us have to tackle every day. However, with a little help and advice, some know-how and imagination and a lot of hard work, I'm certain that everyone is capable of creating a garden they will be proud of, and one that will be the envy of friends and neighbours.

There are obviously many things to think about when planning a garden from scratch, such as how big your block of land is, whether you have children and pets, what the local climate is like, if you enjoy outdoor living and how much time you have to spend on maintenance. All of these factors are important, and they, and many more, will be discussed on the following pages. But first of all you need to know what sort of things you like so you have something to aim for.

Home, garden and general lifestyle magazines are always full of great ideas and photographs to inspire you, so earmark anything that really catches your eye. Your local library is also full of inspiration in the form of gardening books. The same applies to book stores. You can spend hours flicking through the latest releases looking for ideas, and of course you can always buy any book that you think might be useful — just as I hope you did with this one! Television is also a goldmine of design ideas, with one lifestyle show in particular that's clearly leading the field with invaluable tips and ideas.

A walk around your neighbourhood is another way to get ideas, not just for styles of garden, but also to see what plants seem to be happy growing in your particular area. You could even be a devil and knock on someone's door to have a chat about features and plants that take your interest. Most enthusiastic gardeners are only too keen to talk about their pride and joy. Other places for inspiration include garden centres, where not only plants are on sale, but also garden accessories.

All of the sources outlined above will provide you with a wealth of ideas, but you may still feel a bit unsure about the best way to proceed. If so, why not ask a professional to give you a consultation? You'll have to pay, but it's money well spent. If you're impressed by the consultant's approach, you may even decide to get them to draw up a plan. You may even end up asking them to build all or part of your garden, which often happens.

Six of the classic garden styles

Some garden styles have become classics — so well known that almost everyone can conjure up some sort of image in their mind when they hear the name. This is because each of the classic styles has become linked with a set of elements — particular plants, structures and features. A classic style is a good starting point for anyone planning to develop their own garden. You can quickly see which features you need to include to produce a particular effect, and this gives you an insight into what goes into creating a style in the first place. Once you know that, you're well on the way to being able to create a unique style of your own.

Six classic styles — formal English, Australian native, tropical, Japanese, contemporary and Mediterranean — are examined in the pages that follow. Look at the various elements, think about how they go together, and about what appeals to you. You may decide that one of the classic styles is what you really want, in which case your choices will be easy. Alternatively, you may decide that you want the challenge of creating your own style.

I spend a lot of time looking at publications showing the work of other landscapers. It's always interesting to see clever solutions and different ways of doing things.

LET'S MAKE A PLAN

A SENSE OF ORDER

The timeless charm of an English formal garden

There's something peaceful and calming about the sense of order to be found in a well-designed classic English formal garden. The regularity and uniformity of the layout — with paths, hedges and feature focal points working in harmony — imparts a sense of relaxed wellbeing. Symmetry, regularity and balance are the keys.

In many cases, a formal garden accentuates the shape of the house it surrounds, which is often itself also symmetrical. Classic touches are a path lined with box hedging leading up to front steps, and also hedges planted to echo the shape of flanking bay windows, perhaps highlighted by standard roses or topiaries. Any path, fence line or driveway can be framed and highlighted with hedges, often layered, with plants arranged in descending order of size.

This style of garden will work just about anywhere the desired plants will stay healthy, thrive and look good. Apart from box, other good hedging plants to consider include conifers, lilly pilly, viburnum and camellia. For general planting, consider mondo grass, liriope, gardenia, lavender, azalea, magnolia, roses and any white flowering plants. Feature trees can be both deciduous and evergreen. As a finishing touch, a few examples of topiary dotted around the finely clipped lawns will help to emphasise the overall sense of order.

Ornaments and decorations that fit the style include classical urns and statues, the ever-popular lion's heads, symmetrical ponds with a central fountain, plinths, arches and arbours.

Paving materials can be anything you like, as long as they conform to the air of neat and tidy formality. Rustic, second-hand bricks laid in intricate patterns, paths of decomposed granite, pebbles and sandstone will all work. For walls, masonry is popular, whether natural stone or rendered or bagged brick. Less common are non-traditional materials. That's not to say you can't use them, it's just that they're not generally regarded as part of the style associated with the grand country houses of 17th and 18th century England.

It is the way in which plants are constrained and shaped that helps to create the sense of order in a formal garden. The style also exploits subtleties of colour and texture.

PG *recommends*

ALL FOR ONE, ONE FOR ALL

Formal gardens generally involve lots of hedges and borders, and for these to look good it's important that all the plants are healthy and growing at the same rate. Before planting, carry out any soil improvement work, taking care to ensure that the entire bed receives an equal share of organic matter and cultivation. That way you can treat the entire hedge or border as if it were one plant, and this will avoid one or two specimens growing at a different rate to all the others. Making sure all the plants are the same variety seems obvious, but it's surprising how often mistakes are made. There's nothing more annoying than waiting six months for a carefully laid out, symmetrical display to bloom, only to find that two white specimens somehow found their way in among the 98 pink ones. To avoid the embarrassment, buy only from a supplier you trust.

 STYLE CHECK LIST

- Symmetry and balance
- Straight lines and order
- Immaculate lawns
- Clipped hedges
- Decorative urns and statues
- Ponds and fountains
- Arbours and arches
- Topiary and standard roses

LET'S MAKE A PLAN

AUSTRALIAN NATIVE GARDENS

Native species for diversity and variety

Creating an Australian native garden is an exciting project because there are so many options to choose from — rainforest, tropical, coastal, arid, alpine — styles that are as diverse as the Australian landscape itself. It's easiest to create a native garden that uses plants indigenous to your area, but there's nothing stopping you from, say, capturing the feel of the tropics in a Melbourne backyard. You've just got to choose the right plants for your environment.

While there's pleasure to be had in creating the authentic feel of a patch of Australian bush, there's more to the job than just whacking in some natives and hoping for the best. Mixed plantings of various popular native shrubs, such as grevilleas and bottlebrushes, can look spectacular when well planned; they can also look dull and boring when done badly. Also, you don't necessarily have to be a strict purist, since some great effects can be achieved by combining natives with exotics from Asia, Europe and anywhere else.

Also keep in mind that apart from the plants that seem to have become a compulsory part of every native garden, there are literally hundreds of other plants that could be used. Consider, for example, palms such as kentias from Lord Howe Island, bangalows that are indigenous to the east coast, licualas, livistonas, or maybe lilly pillies. The latter are very adaptable, with numerous varieties that are great for hedging and screening, particularly when used in a formal garden. Lilly pillies also work well in a rainforest-style garden. Apart from the trees and shrubs, don't forget all the native grasses, lilies and ferns that can be used, and which offer exciting possibilities.

The choice of landscaping materials will depend on the style you're trying to create, but as a general rule stick to local products to maintain an authentic look and feel. Choices will include large and small river pebbles to line creek beds or other components of water features; native timbers for decks, pergolas and boardwalks; sandstone, limestone, bluestone and granite for paths and walls.

✎ STYLE CHECK LIST

- Predominantly native vegetation
- Plants that attract fauna, especially birds
- Restrained use of exotics
- Indigenous building materials
- Hardwood timber
- Textured brick and stone
- Creek beds

Australia's flora is so rich and varied that native gardeners are spoilt for choice. The style lends itself to informal plantings, although natives can also be used successfully in formal designs.

36

LET'S MAKE A PLAN

TROPICAL STYLE

A corner of paradise in your own backyard

I've always loved tropical gardens, perhaps because I grew up in England where azure-blue seas and palm-fringed beaches were the stuff of winter daydreams. These days, whenever I think of tropical gardens it always brings to mind resorts in Queensland or Southeast Asia, with their thatched pavilions and ornamental pools nestling amid hectares of lush foliage.

If you live in the hot, moist, wet tropics or sub-tropics — roughly anywhere on the coastal strip stretching around the top of Australia from about Coffs Harbour to Broome — then you won't have any trouble creating a tropical garden, since it's the natural style for your area. However, tropical plants don't grow everywhere, so if you live outside the tropical zone and would still like a tropical-style garden, then you will need to adapt a little and work with plants that grow locally but have a tropical look. Choose anything with interesting foliage — the bigger and bolder the better — such as palms and any varieties with large leaves. Unusual flowers and splashes of bright colour will also help to conjure up the authentic feel. Plants such as strelitzias, cannas, phormiums, yuccas, cordylines and bougainvilleas will all handle cooler climates, as will palms such as *Raphis excelsa*, *Phoenix roebelenii*, *Trachycarpus fortunei* and many others. The best way to make a start is with a trip to a local nursery, where staff will be able to help you with advice on the best species.

Timber decks, boardwalks and stonework always work well in tropical surroundings, as, of course, do water features such as ponds and pools. Another great touch is an outdoor cabana or day bed, or better still a thatched pavilion. At night, as with any garden, concealed lighting is always great for setting the mood, with flaming torches for occasions when you want to add a touch of mystery and romance.

If you want to capture the feel of an exotic tropical location — such as Bali, Fiji or Sri Lanka — you can buy artefacts and outdoor furniture from specialist importers and some garden centres.

For many of us, tropical style has been heavily influenced by holidays spent in exotic destinations, particularly in the luxurious resorts of far north Queensland, Bali, Thailand and Fiji.

STYLE CHECK LIST

- Bold, interesting foliage
- Profusion and variety
- Splashes of striking colour
- A 'natural' feel
- Areas for outdoor living
- Water features
- Timber boardwalks
- Exotic artefacts and ornaments

LET'S MAKE A PLAN

AN ORIENTAL RETREAT

Zen and the art of Japanese garden design

The Japanese style of garden design is extremely versatile and can be used anywhere from courtyards and office interiors to large suburban plots. Traditionally, a Japanese garden is seen as containing representations of various natural features of the wider landscape, with rocks signifying mountains; sand or gravel, often raked into patterns to resemble water, representing sea coasts or rivers and streams; and plants grouped to suggest woods and forests.

Zen Buddhism has influenced traditional Japanese garden design, and there are plenty of books covering the subtleties of the art. To be perfectly honest, I don't think I'm qualified to discuss the finer points in any detail, although I certainly appreciate the wonderful results that can be achieved. I love the way that the Japanese really embrace the whole spiritual aspect of garden design and connect with it all on a deeper level than we tend to do in the West. It's not just a case of bunging in a few plants, rocks and a water feature and hoping for the best, although it is possible for anyone to mimic the general effect by using the appropriate materials artfully placed.

Any plants that are native to Japan can be used in a Japanese-style garden, with popular choices being bamboos, mondo grass, maples, camellias, azaleas, conifers, ferns, mosses and grasses. Often plants are pruned into rounded shapes, or grouped together into pleasing, but apparently random arrangements. Don't be fooled, however. The arrangement of plants in a traditional Japanese garden is an art form, and if you're really keen to do it properly you may need to get an expert designer to come and give you some help.

The materials and artefacts often used in Japanese gardens include natural rock; gravels, often white and raked to form patterns; pebbles; slate; timber, either dressed or distressed; bamboo for screens, structures and furniture; traditional lanterns; carved granite sculptures and statues of the Buddha. Bonsai — dwarf trees or shrubs in shallow pots — are always welcome additions.

STYLE CHECK LIST
- Artfully 'natural' design
- Pebbles and gravel, often raked
- Natural rock
- Bamboo, mosses and conifers
- Ponds and streams
- Lanterns and statues
- Screens and fences
- Bonsai

While white gravel, bamboo and stone lanterns are elements most associated with a Japanese garden, the way they are arranged to create a sense of artful informality calls for considerable skill.

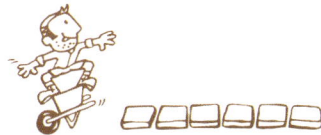

LET'S MAKE A PLAN

CONTEMPORARY STYLE

An eclectic mix of traditional and modern elements

The latest high-tech materials; influences from modern architecture, or sometimes old classics transformed with modern materials; lots of clean lines; multiple examples of the same plants; striking contrasts in colour and texture; clever use of space; trendy furniture; glass; resins; metal; timber; sexy lighting; water and rock: these are all features of contemporary garden design. This eclectic style uses whatever furnishings and materials happen to be in fashion, together with the best that the natural world has to offer, all combined in a funky, modern way. I'm a big fan of this style, and a number of my own gardens would be classed as contemporary.

Functionality is an important feature of this style, as seen in many inner city homes where spaces are small and the interior of the house is extended seamlessly through into the garden, making clever use of the space. One of the main attractions of the style is that it can be adapted to work anywhere in the world by using appropriate local plants and design elements. Particularly good are plants with strong architectural form — such as yuccas or agaves — which bring the hard lines of contemporary house and furniture design to life. I think of this as one of the special characteristics of the style.

Materials can vary enormously, with high-tech products always in favour. If more traditional materials are used they are at least put together in an interesting, modern way. Popular choices are glass — clear, coloured, frosted and sandblasted; metals like copper, stainless steel and mini-corrugated galvanised iron sheets; industrial strength nuts and bolts; rendered masonry and resins. The list could go on, but the important thing to remember is that it's the creative way the materials are used that makes this whole approach work. This can be seen particularly in gardens where ultra-modern and purely natural materials have been combined in clever ways, such as, for example, where a rock rises out of a clean, rendered wall, or where some rustic hardwood hugs an element fashioned from stainless steel or glass.

Interesting combinations of form — particularly natural shapes juxtaposed against the regular shapes of manufactured materials — are one of the hallmarks of contemporary garden design.

PG recommends

SPIKY, SPIKY, ME NOT LIKEY

Wearing protective eyewear while weeding might not seem necessary, but I would strongly advise anyone to do so, especially when weeding close to plants that have sharp, spiky bits. I learnt this lesson the hard way many years ago when removing weeds close to a Canary Island date palm. Those familiar with these palms will know that they have ferocious spikes at the base of the fronds. The sensation I felt as one of these spikes scraped across the surface of my eyeball was very nasty to say the least. I had to drive to the nearest medical centre with one eye shut, not an easy or safe thing to do, but there was no choice. The doctor looked grave, told me I'd had a lucky escape and ordered rest, drops and a patch while the eye healed. Fortunately it did, but it had been a close shave and these days I always wear safety glasses.

STYLE CHECK LIST

- Functionality
- Clever use of space
- Modern, high-tech materials
- Glass and stainless steel
- Traditional and modern combined
- Water features
- Plants with strong forms
- Clever use of colour and texture

LET'S MAKE A PLAN

MEDITERRANEAN TRADITION

A garden to capture that feeling of long, lazy days on the Riviera

Mediterranean-style gardens can work well in Australia because the climate in many areas is similar to that found in southern Europe. The long, dry summers encourage the growth of similar plants, making a Mediterranean-style garden easy to create and maintain.

Geographically, the Mediterranean region covers a large area, taking in parts of southern Europe, North Africa and the Middle East. Around 15 countries have a coast that is washed by the waters of this vast inland sea. However, the areas I'm focusing on for the purposes of this book are the southern parts of France, Italy, Spain and Greece. All of these have a climate that encourages a relaxed lifestyle, with an emphasis on outdoor living and entertaining — a good reason why many Australians will get a lot of pleasure from a Mediterranean-style garden. Courtyards are always popular for creating a feeling of intimacy, while lawns are not so common because of the long dry summers, and the fact that water is often scarce.

Many of the plants to be found in Mediterranean gardens have excellent drought tolerance, which is why they grow so well in many parts of Australia. Olive trees, lavender, rosemary, pencil pines, citrus trees, grape vines, echium and some roses should all thrive.

Popular building materials include terracotta tiles, sandstone paving and masonry for walls, with decorative details provided by tiled mosaics and traditional statues and fountains. Courtyards are often dotted with shrubs and flowers in earthenware pots or large tubs, while a vine-draped pergola provides shade and a space for dining or relaxing in summer. Outdoor furniture is generally simple, with the emphasis on traditional and rustic designs and materials.

Authentic finishing touches include lime-washed walls in earthy shades of cream, ochre and faded red — with splashes of vibrant colour, perhaps blue or red. Inevitably the paved areas will end up with a few splashes of colour too, most likely the red of *vino collapso* which flows in abundance during those long lunches.

✏ STYLE CHECK LIST
- Stone-flagged courtyards
- Earthenware pots
- Terracotta and sandstone
- Pale ochres, creams and pinks
- Fruit and olive trees
- Timber pergolas
- Ponds and fountains
- Geraniums in window boxes

Space for outdoor living is the key to creating a Mediterranean-style garden, providing both shade for summer and sunny spots to enjoy the warmth in winter. Simple, natural materials work best.

LET'S MAKE A PLAN

A STYLE OF YOUR OWN

Planning a garden that will reflect your individual tastes

When it comes to creating your own individual style of garden there are no rules that you have to adhere to. The garden style-police won't suddenly come knocking at your door, although in some areas residents may feel that their local council does a pretty good job of filling that role. What you do is up to you, so have some fun.

Combining various different elements to come up with your own unique style can be very rewarding. You might try mixing natives and exotics, for example, creating differently themed sections in front and back gardens, or sectioning off a courtyard area. Consider all the possibilities, bearing in mind factors such as functionality and the degree of privacy you want. After all it's your garden so you can do what you like with it.

Although there are no rigid rules on what constitutes good design, there are some practical considerations you must take into account, most of which are really just plain common sense. One important point to bear in mind is the selection and grouping of plants with similar environmental requirements. Planting something like lavender, which needs good drainage and generally a quite dry situation, next to native violet that requires plenty of moisture, will result in one or other plant suffering and eventually dying. For information on growing conditions and species that do well together, the staff at nurseries can generally be relied on to provide good advice.

Another important consideration when planning a new garden is maintenance. If you love working in the garden, and have the time, then this won't really be an issue for you. But for anyone with busy work and social schedules, the choices are either a low maintenance garden, or the alternative of paying someone to maintain your garden for you. Obviously, the more maintenance that is required, the more it will cost. Lawns involve a lot of maintenance, simply because they need regular mowing in summer, and also because fertilising, weeding, aerating, top-dressing and watering are necessary if they are to be kept looking good. However, despite all the work involved, many people still regard a lawn as an essential part of any garden. Not only do they provide a place to sit outdoors, but also a place for children and pets to play where adults can keep an eye on them. If low maintenance really is a priority, then consider excluding from the garden any fast-growing hedges, trees that drop leaves constantly, pest- and disease-prone plants and paving or decking surfaces that are difficult to keep clean.

The secret of creating an individual style is to have a plan. Choose and place elements carefully, taking the time to do so and don't stop until you're happy with the result.

PG recommends

YOUR OWN STYLE, NOT NO STYLE

You can, of course, do what you want in your own garden, but that doesn't mean that whatever you do will have 'style'. I remember once discussing a project with some clients who wanted a semi-tropical garden. Fine, except that the husband really seemed to want some roses as well. The discussion became quite heated, and I was trying hard not to laugh at the absurdity of the situation. Fancy arguing with a complete stranger over whether they could have roses in their own garden or not! Still, I think I was right: roses just would have looked wrong in the garden they wanted.

LET'S MAKE A PLAN

COUNTING THE COST

How much do you want to pay?

How much you want to pay — or can afford to pay — is always going to be an important factor in the design of any garden. Grand plans are all very well, but in the end someone has to pay for them. However, this doesn't mean that you have to settle for second best. There are ways of cutting costs and doing things economically that can deliver great results for a modest outlay.

At the top of the scale, in terms of expense, is paying a professional to do the entire job for you from start to finish. For many, this is the way to go. If you can find the right person with the right ideas you will get a first class job done in the shortest time, with no stress and no sweat on your part.

At the other end of the scale is doing the whole job yourself. This is extremely rewarding, provided you have the skills and know-how, and it can save you money, particularly on labour. Bear in mind, however, that although you will save money on labour, you may pay more for some materials and will almost certainly pay more for plants if you buy them at full retail prices. Professionals generally have access to trade prices, and may pass some of the savings on to you. You won't get all the savings, of course, since landscapers, like anyone else in business, need to make a profit.

Between these two extremes there is a whole range of possibilities. You can employ experts to carry out all the skilled work while you do some of the easier jobs, or you can get someone in to do all the heavy lifting while you look after the fiddly, time-consuming details. Almost anything can be arranged, and the best choice will vary according to your skills, your budget and the amount of time you have available.

Before making any decisions on a course of action, start by researching the prices of various materials, such as pavers, soil and plants. Then get quotes for the whole job, as well as the various components that could be sub-contracted. This will help you to create some sort of budget, and will give you a realistic idea of how much landscaping your particular garden is likely to cost.

If you discover that you can't afford your dream garden, don't give up yet. You can always spread the expense by doing the job in stages over a period of time, say a couple of years. Concentrate on getting the most important parts of the design done first, such as some paving or a deck for entertaining. You may also want to get the plants in as soon as possible so they can become established and start growing, and perhaps leave the water feature and lights for later.

Alternatively, you can borrow money against equity, if you haven't done so already for renovations. Shop around and be prepared to compromise if it all seems too much. Remember that a well-designed garden is an investment, not just financially, but also in your quality of life. A garden is a special place to be savoured.

PG *recommends*

PREPARE YOURSELF FOR A SURPRISE

I've done plenty of quotes for clients over the years, and have had varying reactions to the prices I've given. I don't consider myself particularly expensive, but from time to time I've come across clients who were horrified by what they regard as an exorbitant price. It's only after they've had other quotes — often higher than mine — and I've explained in detail what's involved, that they appreciate the scale of the job. There's a lot more to landscaping a garden than you may think, so prepare yourself for a surprise when the landscaper delivers his quote. Remember, if you pay peanuts you get monkeys!

PLUS A Mature, well proportioned gum tree and large kentia palm are important assets to the garden, well worth retaining.

PLUS Deck provides views over the garden and the surrounding area. Faces west, so a good vantage point for viewing sunsets.

MINUS Mulberry tree good for screening and shade in summer, but sheds leaves in winter, revealing view of neighbours' yards.

MINUS Deteriorating old fence and messy lattice will need to be replaced or concealed with some extensive boundary plantings.

LET'S MAKE A PLAN

ANALYSING YOUR SITE

Drawing up a list of pluses and minuses

Site analysis is a crucial part of developing an overall plan. It helps you to identify all the pros and cons of your garden and will assist in providing a more complete understanding of the space you have to deal with.

If you've lived in your present home for some time, you probably think you can identify all your garden's good and bad points off the top of your head, and will be able to write them all down without even bothering to go out of doors. Resist the temptation to do this. When you live with a house or garden for any length of time, you sometimes blank out the things you don't like; they simply recede into the background. We've all experienced that feeling of coming home after a prolonged holiday and being struck by how shabby the kitchen paintwork looks. You will have to make an effort to see the garden with fresh eyes, as if for the very first time. This is why it often helps to have a consultant come in and give you an independent assessment.

Arm yourself with a notebook and pen and walk around the garden — with your partner and children, if you have them — looking at it critically from all angles. Write down a list of the obvious pluses and minuses. On the plus side, perhaps there's an old tree that must be preserved, a great view, or a flat area that would be ideal for a patio. On the minus side, there may be an ugly building across the road that could be screened out, a steep slope that will be hard to deal with, or a rickety old fence that will have to be demolished.

Of course, not all the pluses and minuses will be obvious. You also need to take into account the type of soil you have — poorly-drained heavy clay, for example, will need some remedial work. There may be rocky areas that are hidden just below the surface of the soil, and these can inhibit planting and may affect where you are able to position garden beds. At the same time, if the rock is stable it can work in your favour as something solid that can be paved over, providing stability and reducing the need for roadbase, concrete or crushed rock. One of the most important factors to consider at this stage is access. How easy or difficult will it be to get equipment and materials into the garden?

Once you've finished your survey, prepare a rough sketch of the site, ideally on a sheet of A3 paper. This will give you plenty of room to show all the detail and to make written observations. At this stage, you don't need to draw to scale, but it's a good idea to include measurements showing the lengths of boundaries and the distances between, and heights of all structures. Use arrows to indicate slopes and steps. The important thing is to get everything down on paper clearly and legibly.

LET'S MAKE A PLAN

WORKING WITH THE ENVIRONMENT

The importance of climate, weather and aspect

Weather and orientation are major considerations when planning a garden. If you've lived in your house for a while, you probably know what to expect. But if you're new to an area, you'll need to do some research.

When it comes to the weather, you should know how much sun and rain your area receives at various times of the year, and also what temperature extremes you can expect, particularly whether frosts are likely in winter or prolonged dry spells in summer. Information on any other local climatic factors will also be useful, such as, for example, the fact that there are regular strong winds. You can probably get all the information you need from neighbours or local plant nurseries.

For aspect, you need to know which direction your garden points and also what your latitude is — how far from the equator you are. Latitude is important because this will tell you how high the sun will get in the sky in summer and winter. In Sydney, for example, at latitude 34°S, the sun reaches 32° above the horizon in mid-winter and 79° in mid-summer. Orientation is important because it will tell you where the sun will rise and where it will set. Once you know this, you can work out how shadows will fall across your garden through the year. You don't need to be enormously accurate, but if you have a rough idea it will help in positioning elements such as decks and trees to take advantage of sun and shade at different times of the year.

When it comes to choosing plants, information on climate and weather is obviously essential. If you have a shady garden, you need to think about plants that will be happy growing there. How much rain you get will also influence your choice, particularly now that water is an increasingly scarce commodity.

Wind is an important factor, especially if it carries salt with it from a nearby ocean. There are plenty of plants that thrive in coastal environments, so you can save yourself a lot of time and trouble in the long run by choosing those rather than more sensitive species.

Enter all the climate and aspect information you gather on your site plan (see pages 62–65). Mark a north arrow, as well as arrows to show where the sun rises and sets in summer and winter. Colour in areas that will be in shade for extended periods, and indicate where any prevailing winds come from.

The climate maps opposite give a general idea of the conditions you can expect anywhere in Australia. For more information visit www.bom.gov.au.

 *The show must go on*

We've certainly had our fair share of rainy days when filming *Backyard Blitz**. But there's one occasion that always comes to mind when I think of rain. It was Melbourne, in the middle of winter and freezing cold. It didn't stop raining for the entire two days of filming, and I don't mean just a gentle sprinkle, but a deluge that would have had Noah ordering a second shift on the ark. To make matters worse, the soil in the garden was clay, so the whole site was a quagmire. We were all dressed in thermals and full wet weather gear and still got soaked. It was, without exaggeration, the most uncomfortable two days of my life. But of course it was all worth it to see the look on the owner's face when we revealed his brand new garden to him, even if it was eleven at night, zero degrees and still pouring with rain. There was a smile on every face and a tear in every eye.

AVERAGE ANNUAL RAINFALL

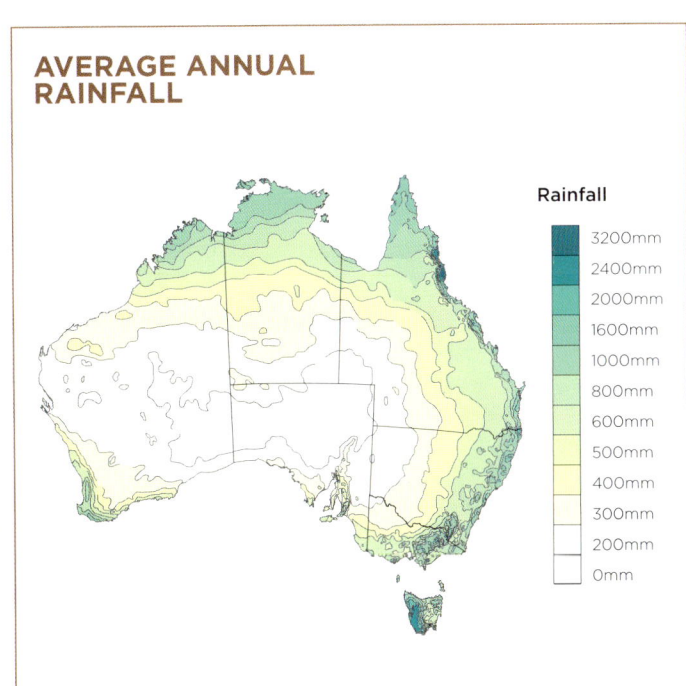

AVERAGE DAILY SUNSHINE HOURS

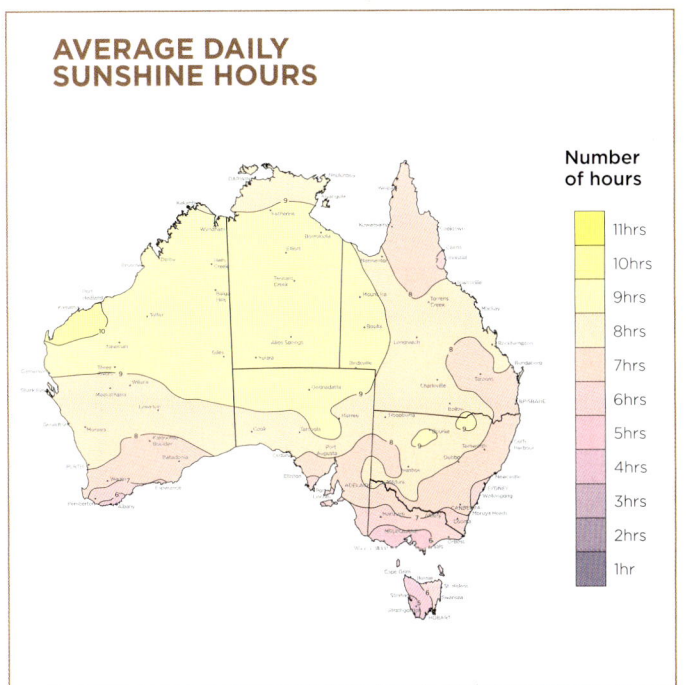

AVERAGE ANNUAL MINIMUM TEMPERATURE

AVERAGE ANNUAL MAXIMUM TEMPERATURE

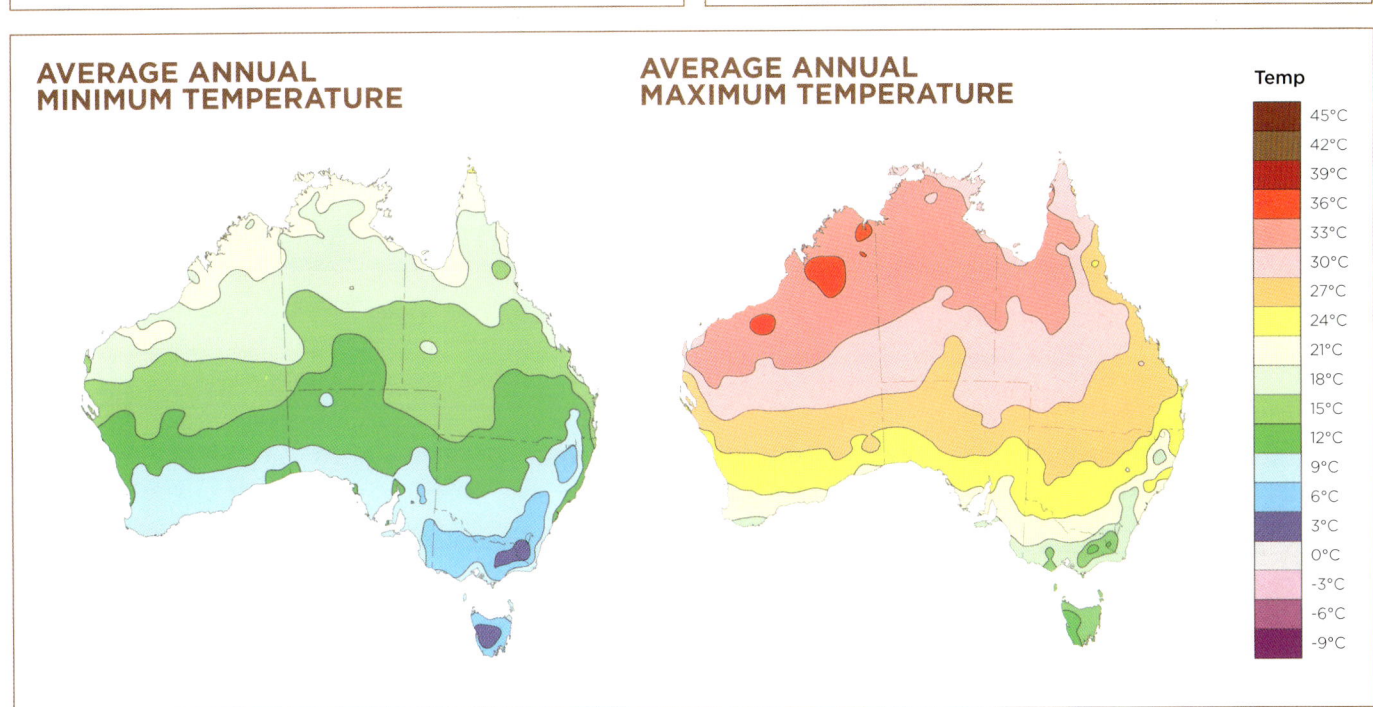

Maps courtesy of Australian Government Bureau of Meteorology.

TYPICAL SOIL DIAGRAM

It can take up to 100 years for a coup of centimetres of topsoil to naturally develop — another reason why soil erosion should be avoided at all costs

Loose organic matter

Decomposed organic matter

Area where true mineral soil and organic matter mix. Generally dark in colour.

Paler area where leaching has removed organic matter.

Unconsolidated fragments of bedrock.

Bedrock

LET'S MAKE A PLAN

DOWN TO EARTH

Soil quality is the key to healthy plants

Soil plays a big part in the wellbeing of plants. It provides varying quantities of nutrients, oxygen and water to root systems, depending on how well balanced it is. To understand soil, you need to know how it is made up, and what functions its components perform. These are:

- Organic material, which is made up of decomposing plants and animals. This improves soil structure and also provides essential nutrients to plants.
- Minerals, which are derived from rocks broken down by weathering. These also provide essential nutrients.
- Air that exists in pockets amongst the soil particles, which provides oxygen to plants and also helps water to penetrate into the soil and drain quickly.
- Water, which contains dissolved nutrients essential to growth in a form that is accessible to plant roots.
- Living organisms, which range from micro-organisms, some of which help root growth, to earthworms, which improve soil structure by creating air spaces and breaking down organic matter.

Soils vary a lot because they have very different proportions of these components, arranged in different ways, and this in turn affects the way they behave. It is important to know what sort of soil you have since this will tell you what you may need to do to improve it to create the best environment for plants (see page 112).

To analyse your soil, you need to have a good look at it. Dig some up and feel it in your hand. Make note of what colour it is, and how the colour changes according to depth. Now carry out a 'ribbon' test to find out what sort and size of particles your soil contains — sand, silt, or clay — and in what proportion. Pick up a handful of soil and work it together, adding a little water at a time until it is moist. Note the different textures. Sandy particles are coarse and easy to feel, while the smoother particles are clay, silt and humus. Now squeeze the sample to try and form it into a ball. If the soil contains a fair proportion of clay and humus this should be easy, but if it contains a lot of sand it will keep breaking up. Next, gently squeeze the ball to lengthen it into a strip or 'ribbon', and see how long you can make it before it starts to break up. The more clay the soil contains, the longer the ribbon will get and the more plastic it will feel to the touch. Soil ribbons can range in length from about 5mm to 75mm, with sandy soils being shortest, loamy soils in the middle and clay soils the longest.

While soils that contain a lot of clay or sand may need work to improve them, loamy soils are generally the ideal. These contain a mixture of sand, silt, organic matter and clay in varying quantities, providing the right balance of nutrients, oxygen, water and drainage.

Now look at the soil structure — the way that the different soil particles are arranged, the aggregates they form and the spaces there are between them. Well-structured soils will have aggregates firmly bound together, which will help to maintain the structure when the soil is wet, but at the same time be loose enough for plant roots to move freely in search of water and nutrients without using too much of the energy that is needed for growth and development above ground. Well-structured soil will also have plenty of air spaces to provide oxygen, allow water to penetrate and facilitate drainage. Poorly structured soils are often hard and compacted, will not accept water easily, have few air spaces and are subject to erosion.

Last, but not least, you need to know how acid or alkaline your soil is. This is measured on what is known as a pH scale, which ranges from 0 to 14. Lower numbers indicate acid soils, higher numbers alkaline soils, while neutral soils are somewhere in the middle. A pH measurement is important because some plants like slightly more acid soils while others prefer slightly more alkaline conditions. For most general purposes, it is best to have a soil with a pH of between 5.5 and 7.5.

LET'S MAKE A PLAN

CHILDREN AND PETS

Planning for a child- and pet-friendly garden

If you have children or pets — or are planning to acquire some in the near future — you obviously need to consider them in the planning stages of any new garden. Kids and dogs like to have plenty of space to run around in, and the most obvious choice for a play area is a lawn, which is both soft and safe. However, lawns need a fair amount of maintenance, especially if they get a lot of wear. If you're planning a new lawn, choose a hardy species if possible, but most importantly one that will thrive in your climate and situation. Cats are less of a problem, but they do tend to stray, especially at night. If your garden borders a busy road, you may want to take some measures to make the danger area harder to reach if you don't want moggy to become yet another accident statistic.

A play area is great for small kids. Equipped with the standard sand pit, swing, slide and climbing frame it will keep them amused, active and outdoors for hours at a time. When planning a play area, give careful thought to its location. It will need some shade, and you might want to shield it from entertainment areas, but you will certainly want it to be clearly visible from the house so that small children using it can be constantly supervised. This generally means that it has to be visible from the kitchen, although individual circumstances will vary. For kids of the right age, a tree house is always popular — provided you have a suitable tree in the garden. Once again, however, make sure it can be seen from the house.

With sand pits, you should have a cover for times when they're not in use. Cats will often use the soft sand as a toilet, which is a health risk, and spiders and other creepy-crawlies are often unwelcome guests, especially if the area is not used for any length of time.

In fact, safety must always be the major consideration when planning a pet- and child-friendly garden. A pool must be fenced by law, but kids have a way of getting around things, so as with play areas, you need to be able to keep an eye on any area of water from the house. Also avoid plants with sharp, spiky foliage. Walls of any height are potential accident zones, so consider making them inaccessible by planting hedges close by, or by blocking off the danger with closely spaced planters.

PG *recommends*

NO MORE NASTY SURPRISES

Dogs are great companions, but they can create havoc in the garden. Digging and chewing are a nuisance, but pooing, can be a health hazard. Dog poo and the inevitable urine that accompanies it, can leave a nasty, long-lasting scar on the lawn. Good training is the key. You must teach your animal to do its business in some less conspicuous part of the garden you have chosen for that purpose. Once that's achieved, you might want to invest in a doggy toilet. This is a great system that consists of a small container with a lid that you bury in a pit in the ground. Dog poo is placed into it where it gets broken down, the residue seeping safely and harmlessly into the surrounding soil. Of course, the trick is to get your dog to replace the lid after use!

LET'S MAKE A PLAN

WATER FEATURES

Decorative and practical additions that will enhance any garden

Water — in the form of a pond, stream, cascade or swimming pool — can be both a decorative and a practical addition to any garden. A tranquil pond can create an air of meditative calm, while a swimming pool provides a focal point around which outdoor activities naturally concentrate. While all types of water features have been popular in recent years, restrictions on water use and alarm at dwindling supplies in many Australian cities has started to change attitudes towards them.

I'm a great fan of water features, and wouldn't discourage them entirely, but I do think that every garden owner must do their best to encourage sensible water use and minimise waste. We have to acknowledge that some features can be more wasteful than others, with leaks and evaporation being the main culprits. My recommendation is to avoid any features that allow water to splash or be blown around by wind so that it ends up on adjoining paths, and not in the reservoir where it's supposed be. You'll be aware of excessive water use if you have to top the reservoir up by hand, but probably not if it's topped up automatically by a float valve. It's therefore important to make a conscious effort to gauge usage by keeping an eye on water levels, particularly on hot days when evaporation is at its greatest. If you are using excessive amounts of water, take measures to cut down. Even minor changes, such as providing some additional shade on summer days can make an appreciable difference.

Backyard swimming pools are always going to be popular in a hot country like Australia. If you want one — either now or a bit later down the track — you will have to plan for it well in advance. Such a large object is obviously going to have a major impact on any garden design. You're hardly going to move a pool once it's in place, so you need to get its location right the first time round. My advice is to call on the services of a professional pool building company to help make the decision. They will be able to offer you a lot of practical advice on size, type, position and surrounds, often based on years of experience. If you like what they have to say, ask them to give you a quote. After all, not many people are prepared to tackle the job of building a pool on their own.

Safety is a priority with any body of water, especially where the children are concerned. In the case of a swimming pool, you will need to install a fence, so allow for that in the budget. Costs can vary considerably, depending on what you want. The standard pool fence is a powder-coated metal construction, but you can get seamless glass fences that are almost invisible. There are usually laws governing fencing around water. Some require a fence for any body of water over 300mm in depth, for example. Check with council before making a decision. However, fences aren't the only answer when it comes to water safety. A galvanized or stainless steel grill over the top of a reservoir — covered with decorative pebbles — will also stop children from falling in.

Other factors to consider when planning a water feature are the extent of any plumbing and electrical work that needs to be done, including items such as pump controls and lighting. All plumbing and electrical work must be done by a licensed professional — by law — and some complicated water features call for a lot of highly skilled work, which will not be cheap. Consider the costs carefully, and if the installation you are contemplating starts to look too expensive, give some thought to simpler, less ambitious options. There are plenty of prefabricated units available from garden suppliers, and these are generally easy to set up.

The vital factor with any water feature is safety. Even if you don't have children yourself, there may still be visitors to your garden and their wellbeing is your responsibility.

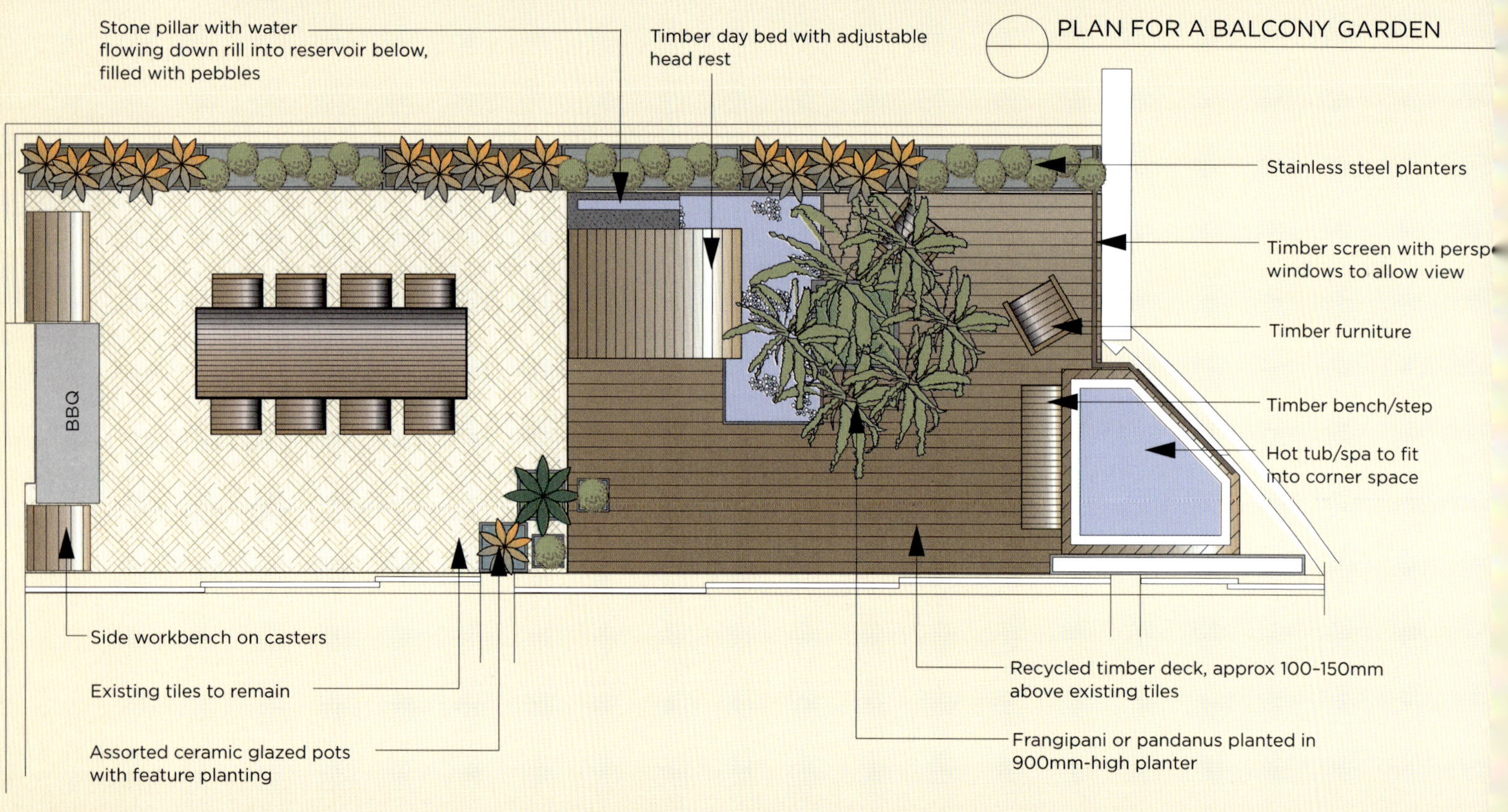

PLAN FOR A BALCONY GARDEN

Stone pillar with water flowing down rill into reservoir below, filled with pebbles

Timber day bed with adjustable head rest

Stainless steel planters

Timber screen with perspex windows to allow view

Timber furniture

Timber bench/step

Hot tub/spa to fit into corner space

Side workbench on casters

Existing tiles to remain

Assorted ceramic glazed pots with feature planting

Recycled timber deck, approx 100–150mm above existing tiles

Frangipani or pandanus planted in 900mm-high planter

LET'S MAKE A PLAN

SMALL GARDENS

Tips and tricks for making the most of small spaces

With more people moving into units and townhouses, courtyard and small gardens are increasingly popular. However, lack of space means that coming up with an interesting design can be a challenge.

With courtyards, a popular and attractive idea is to make the garden an extension of the house, with large — preferably glass — doors opening onto it as seamlessly as possible, giving a sense of space. Pavers or tiles that match those inside will reinforce the illusion, as will clever exterior lighting.

Raised planter boxes placed against walls can help to make the best use of space, as can seats mounted on walls, which will reduce the need for bulky furniture. To prevent the garden from developing a square, boxy feel, build curved or stepping-down walls or garden beds. Alternatively, you can opt for movable pots and troughs, which will give you the freedom to change the layout from time to time. You can add character to walls by cladding them with stone, timber slats or a decorative render, and a painted feature wall or espaliered plants will create a focus of interest.

The one important rule is to keep things simple, since this will help to make the space seem larger. Also, don't use too many colours or materials, and certainly not too many different types of plants. When choosing plants, find out how large they will grow, and avoid species that will eventually outgrow their position.

Many of the courtyard principles also apply to small gardens, although the extra space means you have more options. If the space is large enough you may even be able to consider a small lawn, which is great if you have kids or pets, and easy to maintain if only small.

Balconies are the last word in small gardens. If your balcony is really tiny, there is probably little more you can do other than install some planters — take care to choose species to go in them that will thrive in the conditions on your particular balcony — a few pieces of well-designed outdoor furniture and a BBQ. As before, simplicity and functionality are the keys. Avoid clutter at all costs. With larger balconies, some of the courtyard ideas described above may be possible. Decking will also change the feeling of a balcony, but take care that you don't raise the floor to unsafe levels.

Tight squeeze

People often ask me what's the hardest garden I've ever worked on, often assuming that the bigger it is the harder it is to work on. But one garden that really sticks in my mind is a tiny one we made-over on *Backyard Blitz**. It was an inner-city courtyard that you could hardly swing a cat in — not that we tried to of course, but if we had, it would have been tough! It was tight with the four of us in there, but when we added the rest of the production team — camera men, sound recordists and producers — there was literally no room to move. On top of this, the only access to the garden was through the house, so we were going backwards and forwards with barrow loads of soil and lengths of timber with someone always in our way. Getting the job done was challenging to say the least — it was absolute mayhem from start to finish but it was also very funny, which certainly made for good TV!

The principles of good garden design remain the same however large or small the available space. Have a clear idea of what you want from the garden and then plan accordingly. It's always a good idea to have a focal point, even a very modest one.

LET'S MAKE A PLAN

PUTTING IT ON PAPER

Clear, carefully drawn plans will make your job easier

Once you've decided what you want in your new garden you are ready to start preparing a plan. This process will take you from the first rough sketches that you drew up while you were analysing your site (pages 50–55) to a full working drawing that is accurate enough to build from. Your final drawing should be to scale, have all the necessary detail and be easy to understand. Take plenty of time to play around with your ideas in sketch form to get a better idea of what you want and how it looks down on paper. Take your sketches out into the garden frequently and try to visualise what your pencil marks will look like when they're converted to actual walls, paths and garden beds. If you've never done anything like this before, take your time; don't rush through this stage just because you're keen to start digging. Improvements and alterations that are easy to make on paper may take days of backbreaking labour to achieve in real life.

The first stage is to convert your initial sketches to scale. Some people find graph paper easy to use for this and most art supply stores stock a range of types and sizes. However you do it, it is important that the piece of paper you use is large enough — perhaps A3 or A2 — to show all the necessary information clearly. There is nothing more confusing than trying to work from a plan where the information is crammed in and hard to read. If in doubt, get a larger piece of paper.

The idea of drawing something to scale is that measurements on the plan are in the same proportion as they are on the ground. This enables you to convert from one to the other accurately and quickly. Typical scale sizes for residential landscaping plans are 1:50 and 1:100. At a scale of 1:50, 1mm on the plan represents 50mm on the ground, and at a scale of 1:100, 1mm on the plan represents 100mm on the ground. A scale ruler is an essential tool for this sort of work — you can buy one quite cheaply from an art supply store — and it will also be useful when you start work and need to measure from the plan to get the on-ground dimensions.

To draw up the plans, start with the boundaries, the house and any fixed structures. Then gradually add all the other details. Be patient, as this can take some time, especially if you've never done it before. After all, it doesn't matter if you have to re-draw the plan ten times before you're satisfied with the result. Colour in walls, paths and vegetation if you want to, as this can sometimes make it easier to see where structures start and finish. Professionally produced plans can be works of art, worth framing and hanging on a wall.

Allow plenty of room around the plan to write down details of measurements, as well as descriptions of the various elements and structures. Also allow room for a plant schedule — the names, quantities and sizes of plants — and construction notes. Alternatively, this information can be written on a separate piece of paper. Once the plans are complete and you're happy with the results, have them, and any additional notes, laminated as protection against rain and dirt when you take them out into the garden.

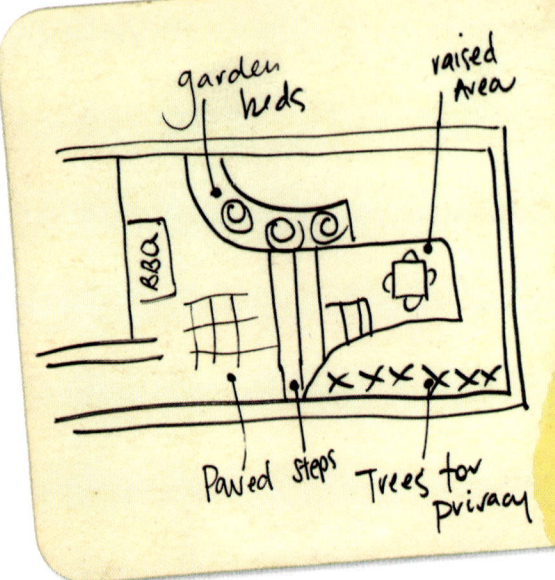

First rough sketch

The first stage in the evolution of a landscaping plan involves putting all the information gathered into one single scale drawing, possibly on graph paper. From there it can be gradually refined and added to.

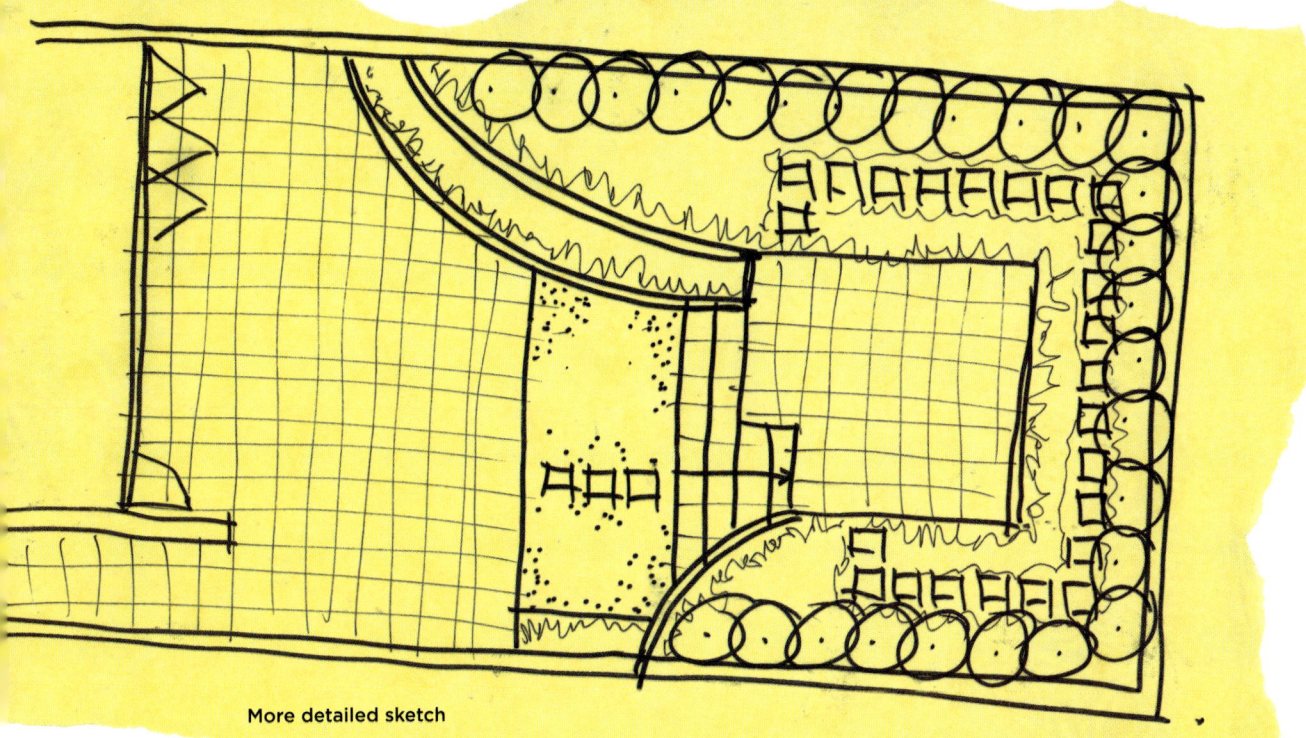

More detailed sketch

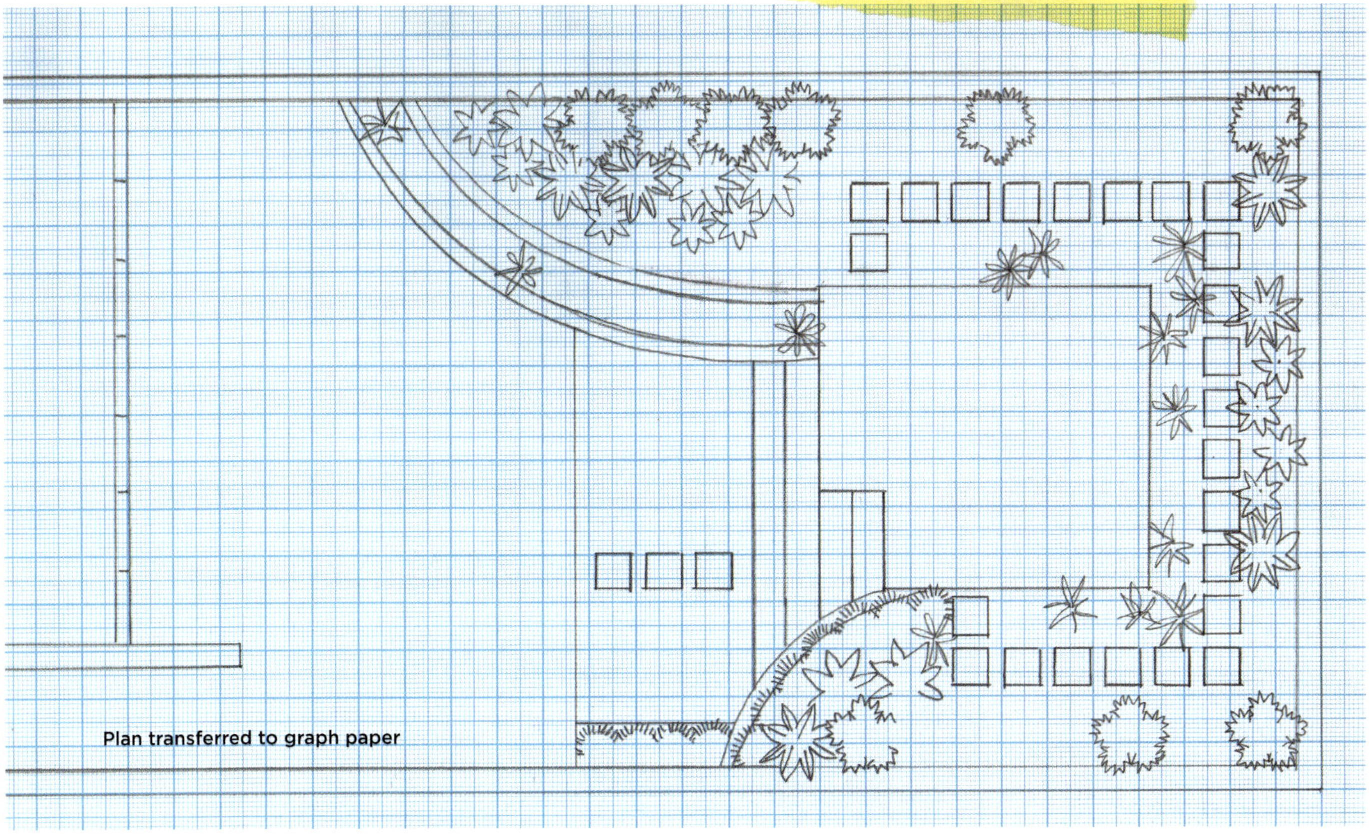

Plan transferred to graph paper

PRODUCING AND REFINING A PLAN

The final working plan, complete with cross section details

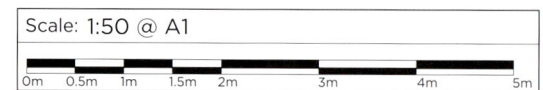

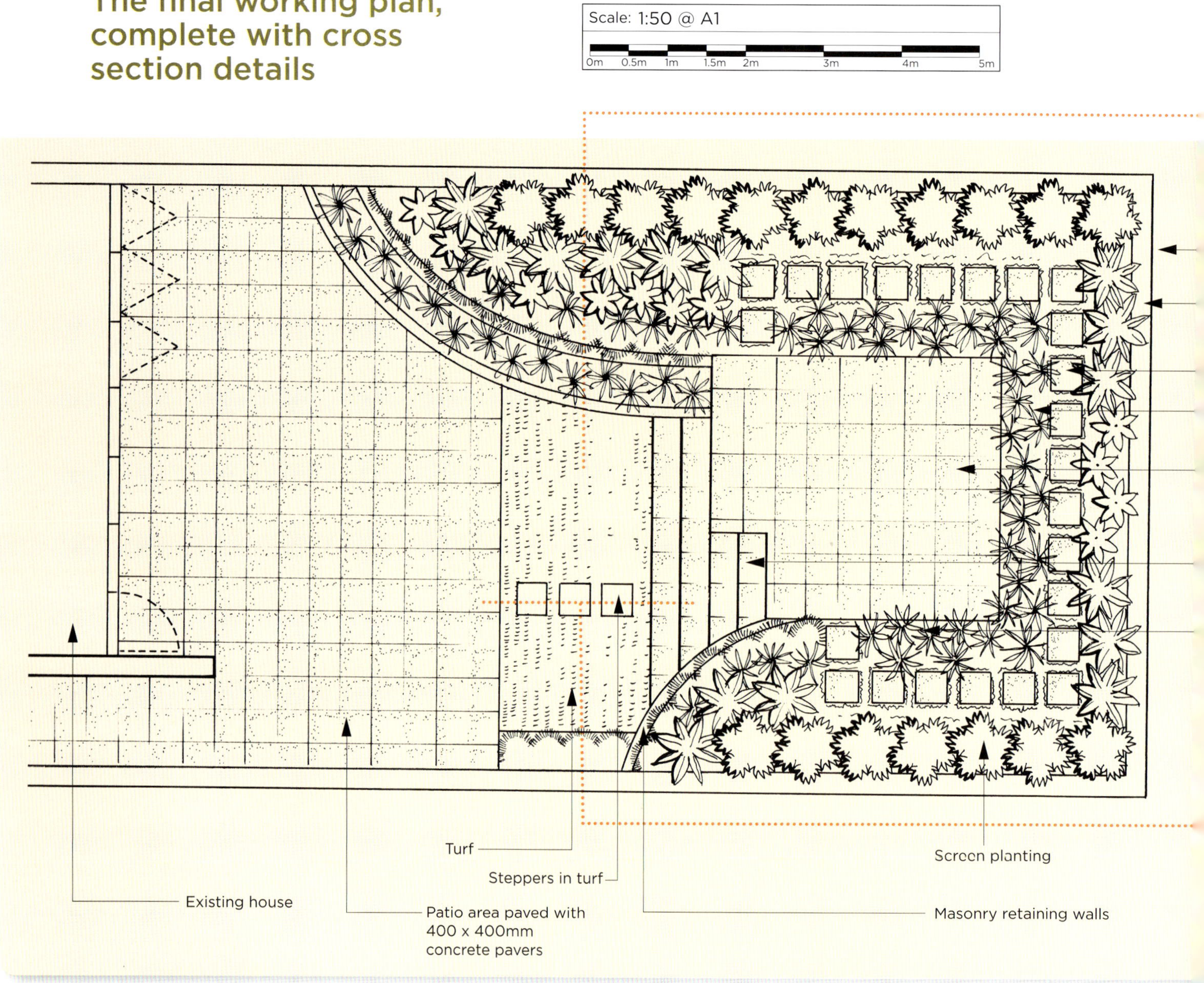
Scale: 1:50 @ A1
0m
0.5m
1m
1.5m
2m
3m
4m
5m
Turf
Steppers in turf
Existing house
Patio area paved with 400 x 400mm concrete pavers
Screen planting
Masonry retaining walls

- Existing boundary wall

- Small palms

- 400 x 400mm Steppers in garden bed

- Ornamental grasses and groundcovers

- Raised entertaining area paved with 400 x 400mm concrete pavers

- Steps

- Raised garden beds

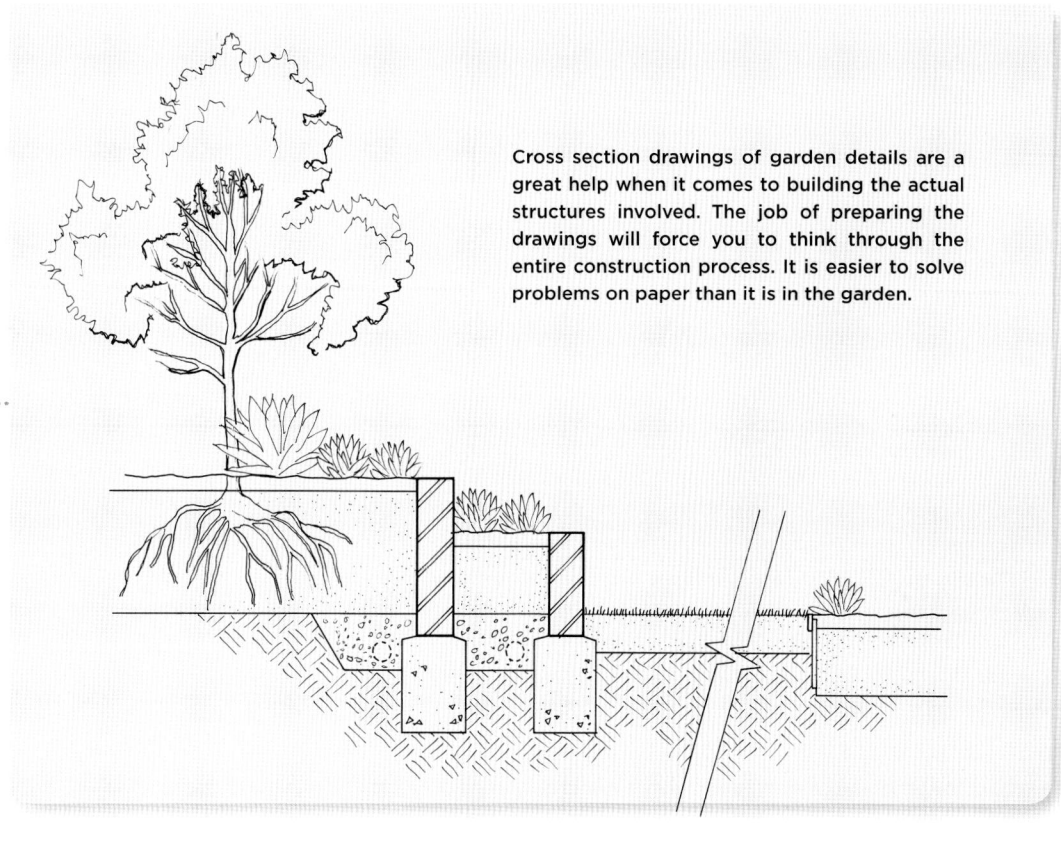

Cross section drawings of garden details are a great help when it comes to building the actual structures involved. The job of preparing the drawings will force you to think through the entire construction process. It is easier to solve problems on paper than it is in the garden.

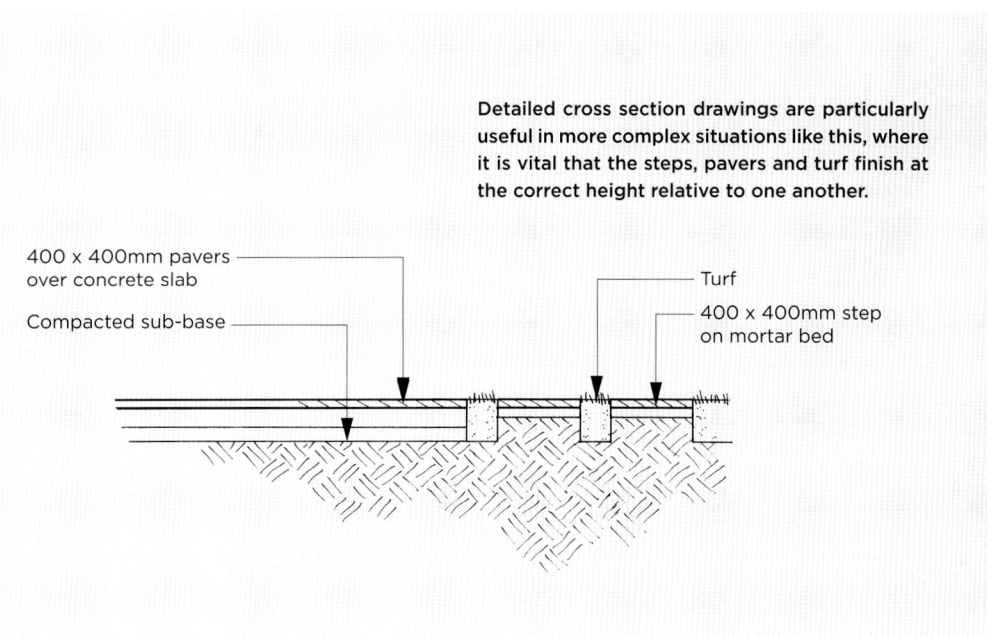

Detailed cross section drawings are particularly useful in more complex situations like this, where it is vital that the steps, pavers and turf finish at the correct height relative to one another.

400 x 400mm pavers over concrete slab

Compacted sub-base

Turf

400 x 400mm step on mortar bed

LET'S MAKE A PLAN

DEALING WITH RED TAPE

Staying legal takes time and trouble, but it's for your own protection

Dealing with all the inevitable red tape before you start a landscaping project can be tedious, but it's something you can't avoid. You don't want to end up on the wrong end of a nasty legal dispute that can cost you not only stress and sleepless nights, but also lots of money. How much red tape you encounter will depend on where you live and the type of project you're planning, but here are some of the areas you may need to consider.

Local councils have a range of rules and regulations covering, among other things, structures, boundary and retaining walls, fences, pergolas, pollution prevention, tree preservation orders and colour schemes. Also, you may be required to buy a permit for skip bins, and some deliveries, without which you may be fined. Requirements will vary according to where you live, so make sure you pay a visit to the council as soon as you have some idea of what you intend to do. This is especially important with some larger projects, where you may need a development application, and that can take time to approve. Once you know what the regulations are, you may even decide to make changes to the size and scope of your project so that permissions are easier to obtain.

Depending on the size of the job and the way it is undertaken, you may need to obtain certain types of insurance, such as public risk/liability, home owner's warranty, workers' compensation and building materials cover. Each job will be different, and the best way to be sure you have the necessary cover is to contact an insurance broker and tell them what you plan to do. Contractors should have certain types of insurance themselves already — which you must remember to check before they start work — so there's no need to double up. Don't risk being uninsured.

It's up to you to protect the environment and to prevent any form of pollution. State environment protection authorities, along with local councils, are the governing authorities in this area, and they also have responsibility for waste management and disposal of hazardous materials. The authorities have the power to come down hard on offenders and they will do so if necessary. Contact them for more information on how you may be affected.

PG *recommends*

LIFE JUST AIN'T FAIR

I got a nasty reminder of how careful you need to be where pollution is concerned a couple of years ago. We were halfway through a big job that involved walls, drainage, plants and of course paving. I had to be away from the site for a few days and left everything in the capable hands of the boys working for me. They were using a brick saw to cut in paving, and had it set up correctly with sediment waste water directed into a silt trap bin to avoid run-off into the gutter, as well as hay bales to act as a sediment trap. Everything seemed OK until the council rangers saw something they didn't like — dirty water running down the gutter past our site. I was slapped with an $850 fine, even though the boys pointed out that two doors up the street there was another construction site where a bobcat was running backwards and forwards from the property to a truck parked on the street, loading excavated soil. We all knew that bobcat was the culprit, but council refused to back down, despite a solicitor's letter, and I just had to pay up. The lesson is you can't be too careful where sediment control and pollution are concerned. Don't pollute, but don't put yourself in a position where anyone can even suspect you of polluting either.

It's tempting to cut corners with the legal paperwork, but I would strongly advise against it. Councils not only have the power to issues fines, but also to order the demolition of unapproved structures.

DOING THE HARD GRAFT

THIS IS WHERE THE FUN STARTS — WHEN YOU CAN GET OUT THERE AND DO SOME REAL WORK. NOW ALL THOSE WEEKS OF CAREFUL PLANNING AND RESEARCH CAN BEGIN TO TAKE PHYSICAL SHAPE IN THE FORM OF ACTUAL WALLS, PATHS, PATIOS, DECKS AND GARDEN BEDS ON THE GROUND.

4

DOING THE HARD GRAFT

PLANNING AHEAD

A detailed schedule of work will keep things running smoothly

There's just a little more planning to be done before you start on the real work. Whatever size of job you're doing, it's important to draw up a schedule of works so you know what order things are going to be done in. Once you have this, you can plan ahead and organise things like the delivery of materials and skip bins, and dates when tradespeople are to start work. Having this written down on paper in front of you helps with the smooth running of a job. Most of the planning is commonsense, but the steps are often not obvious until you've had a chance to think the whole job through.

If different tradespeople are to work on your garden — such as concreters, carpenters and paving contractors, for example — talk to them about timing and availability, especially if more than one trade has to be on the site at the same time, or in some particular order. Sometimes, one job needs to be finished before another can be started. There's nothing worse than doing things in the wrong order, except maybe doing things twice!

The situation will vary from site to site, of course, but the checklist opposite would be a fairly typical order of work for a landscaping project.

BETTER SAFE THAN SORRY Before starting work, take a few minutes to identify and note down any hazards that could cause an accident. Even if it's your property, and you're very familiar with it, don't skip this step. Hazards might include: tripping points; slippery surfaces; jutting branches or other objects at head height (or even worse, at groin height); uneven ground that could twist an ankle; overhead power lines and underground services. It's particularly important that you locate any water pipes, power cables, stormwater and sewer pipes, gas pipes, telephone and cable TV lines. Contact the Dial Before You Dig service — phone: 1100; internet: www.dialbeforeyoudig.com.au — and the relevant service supplier should be able to help. However, to be on the safe side, you may still want to have a poke around yourself.

Planning checklist...

- [ ] Obtain all the necessary council permissions and permits.
- [ ] Familiarise yourself and others with any potential hazards or risks and write them down.
- [ ] Identify what is staying and what is not; protect trees and structures from potential damage; identify a safe storage area for materials when they are delivered, where they will not be in the way of other work.
- [ ] Install any sediment or erosion control and safety fences; protect stormwater pits and drains with silt traps; put out witches hats; make the site safe; have tarpaulins at hand to put down under materials, and over anything that needs protecting if it starts to rain.
- [ ] Remove anything that is not wanted; if necessary, spray grass or persistent weeds and allow herbicide time to take effect (2–3 weeks in warm weather, longer in cool weather).
- [ ] Transfer details from the plan to the ground, positioning the various elements and features.
- [ ] Establish finished heights, excavation depths, falls and slopes.
- [ ] Excavate footings for paving and walls; dig trenches for services; make level changes; cut and fill; grade slopes.
- [ ] Install services, such as stormwater pipes, sub-surface drainage, gas, electrical and access conduits for services such as lighting and irrigation.
- [ ] Build and install structures such as walls, paving, concrete slabs, steps, decks, pergolas, fences and edging.
- [ ] Carry out any soil improvements, including importing and placing new soil.
- [ ] Have any specialist work carried out, such as irrigation, lighting and sealing.
- [ ] Complete soft works, such as planting, mulching and turfing.

DOING THE HARD GRAFT

CLEARING THE SITE

Getting back to basics

Before you lift a finger, make sure you know where all existing underground services in the garden are located. The Dial Before You Dig service (see page 70) will help you locate the main services, but they won't know everything. For the complete picture, you'll have to carry out some investigation of your own. This involves careful digging around downpipes, drains and taps to see which direction pipes are laid, and also checking with contractors or previous owners. The more you find out, the less likely you are to get a nasty surprise. This is especially important if you're planning to use earthmoving machinery, which can cause a lot of expensive damage very quickly.

It is also important to make sure you have any waste disposal organised well in advance, such as skip bins (with permits if required), a truck or a trailer. Thorough planning is especially important if you are using a machine such as a bobcat. Talk to the operator well before starting work so you have an idea how much waste there will be, where to put bins if you are using them, or whether the contractor is going to take the waste away in a truck. Don't forget you may need to organise street space the night before bins are delivered.

Clearing the site involves just that, removing anything that's not wanted, by hand, by machine, or as is often the case, using a bit of both. I'm used to doing the clear out as quickly and efficiently as possible, but it can be done gradually over a period of time. If there's a lot of ground clearing and excavation to do, a bobcat and operator are worth their weight in gold. It's amazing how much they can do in a day, but they do need careful supervision, and you need to make sure you have the name and number of a plumber and electrician on hand, just in case. For smaller jobs, you can probably do everything by hand, and may even be able to bribe a couple of mates to help with the promise of a BBQ and refreshments at the end of the day. Whichever way you decide to work, always remember safety first!

If you do decide to hire a machine, you can maximise the use you get out of it by having bulk materials — such as soil, blue metal and roadbase — delivered at the appropriate time. Then the machine can be used to transport them to where they are needed in a fraction of the time it would take to do the job by hand. This is another good reason why you should talk to the operator well in advance.

Nige says...

THE MIRACLE OF THE NINE BINS It's amazing how skip bins can miraculously fill themselves up overnight, particularly in the city. Ant, Jonno and I once did a job in Surry Hills, an inner Sydney suburb, where obliging neighbours helped us to fill nine 5-cubic-metre bins in only a month or so. When I originally estimated the job I'd reckoned that only five bins would be needed! Of course parking was at a premium in the inner city, and every time a new bin arrived I'd watch with admiration as the operator painstakingly tickled the full bin out — while blocking the entire street of course, so no pressure — and sandwiched the new bin into place between the brand new BMWs and Mercs! Anyway, I didn't lose too much sleep over the cars, since I figured it was their owners' old books, furniture and mattresses that I was paying to have carted away to the tip.

Machines certainly speed up the job of clearing a site. Work that might take a week or more if done by hand can be finished in a single day. Although the machines may look easy to operate, don't be fooled. This is a job for professionals, so when hiring always arrange for the machine to come with a skilled operator.

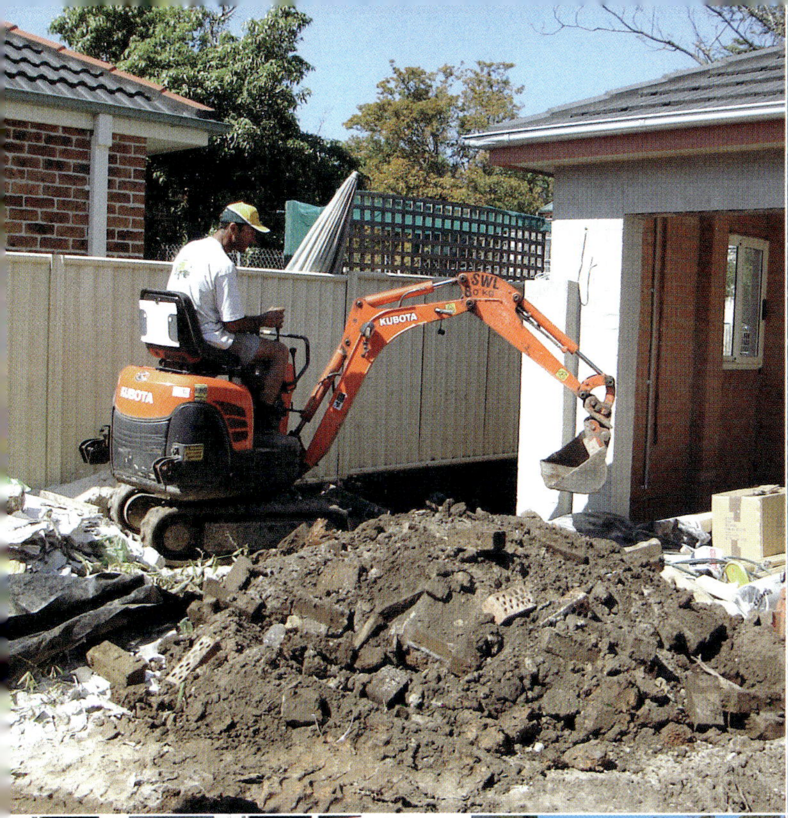
KUBOTA
SWL

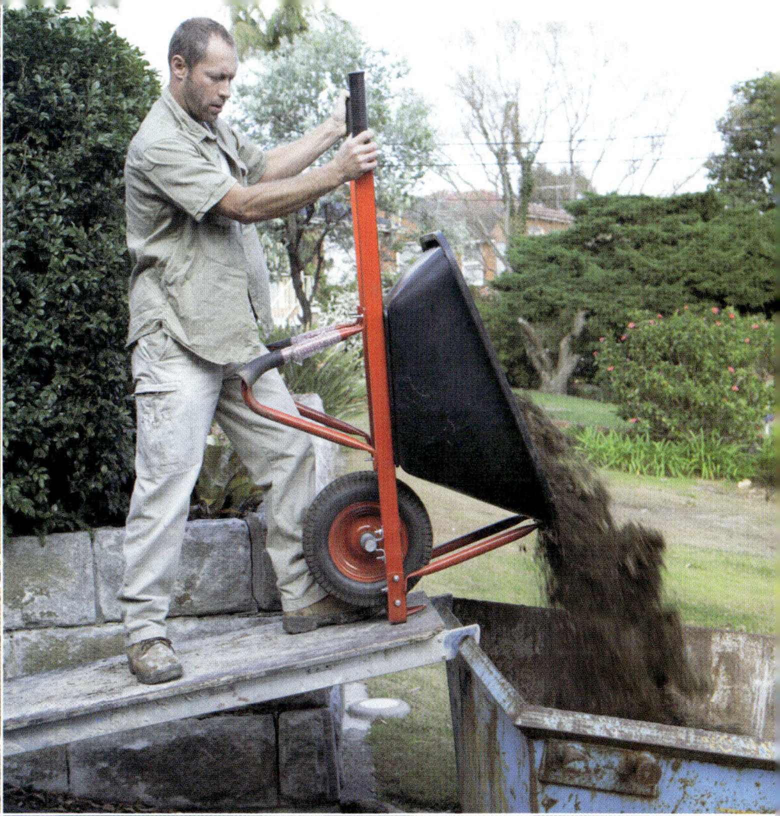

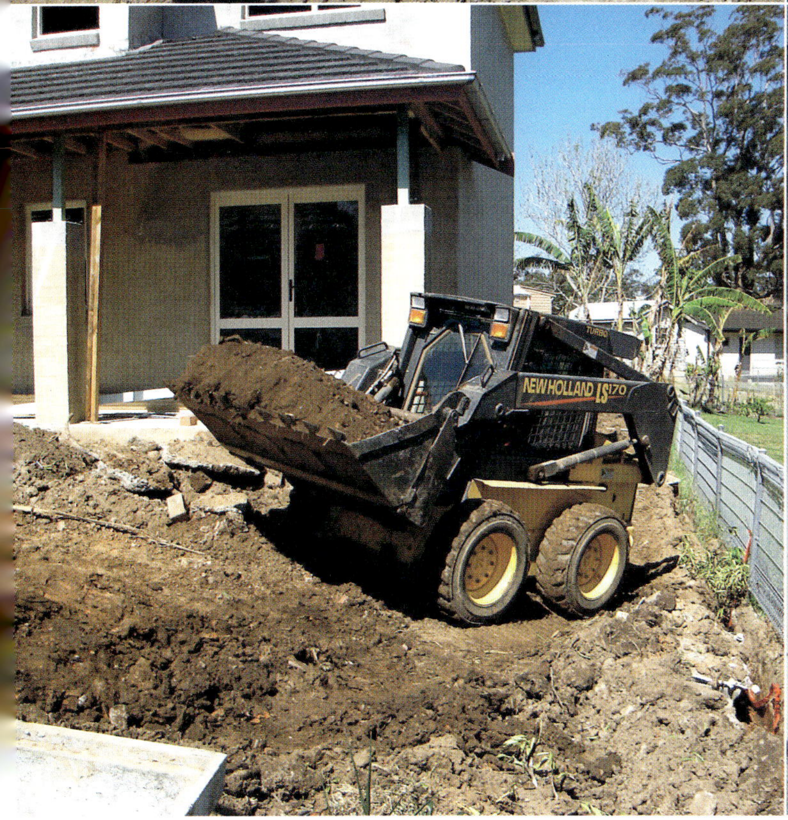
NEW HOLLAND LS170

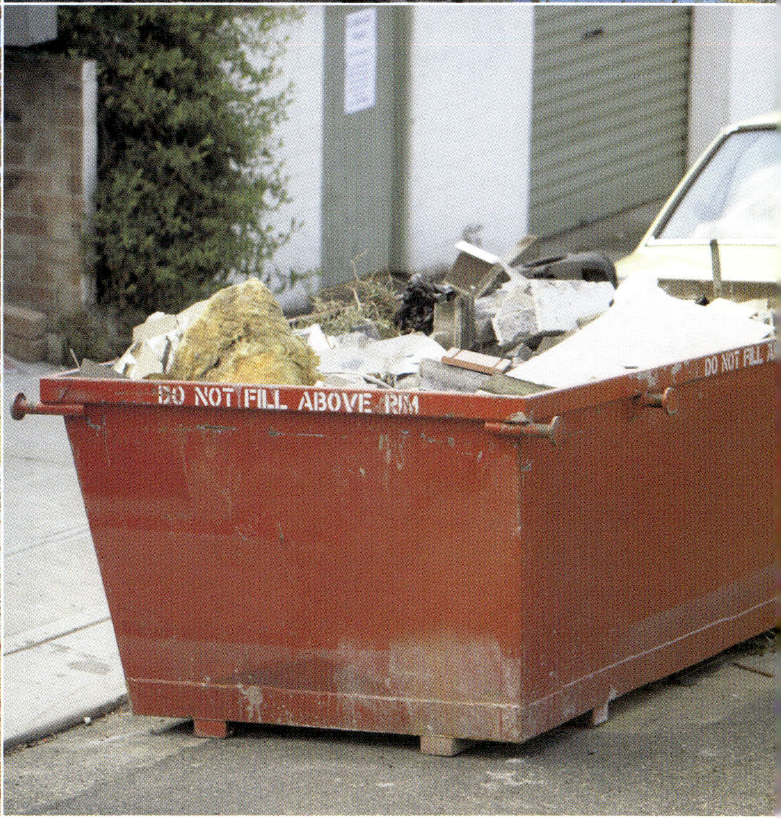
DO NOT FILL ABOVE RIM

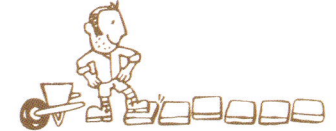

DOING THE HARD GRAFT

FROM THE PLAN TO THE GARDEN

Setting-out

Now that all the preparations have been completed, you are finally ready to transfer the details of your plan to the garden. This job is done in three stages. First of all you must rough-out the plan, marking the approximate position of elements in your garden; next you establish levels so you know the height of the various elements (page 76); and finally you complete a detailed, accurate set-out, showing exactly where things will be (page 78). To complete the rough set-out, measure approximately where the elements will be, and then outline them with chalk, paint or lengths of timber laid on the ground. The tools and equipment you need for the set-out — with instuctions for use — are shown below and right.

1. Builder's square 2. String line
3. Tape measure

STRAIGHT LINES

Walls, paving and paths generally follow straight lines. The basic tool for laying them out is a string line — a roll of strong nylon cord. To mark lines on the ground you can use either a chalk line or a spray can marker.

STRING LINES must be kept as taut as possible. You can anchor the ends to nails fixed to pegs that have been driven into the ground, or by using a few bricks stacked up.

MARK THE CORNERS of a trench or area of paving by extending the string lines beyond the work area and fixing them to pegs where they won't be disturbed by digging.

A CHALK LINE can mark lines on any flat surface. Stretch it tightly across the surface, pinch it between two fingers, raise it a short distance and let it 'snap' back.

SPRAY CAN MARKERS are handy for roughly indicating the edges of areas where soil is to be excavated, or places where various structures are to be positioned.

STAYING SQUARE

There are two basic tools — a builder's square (often used with an extension) and a timber square — that will help you to produce accurate 90° angles on the ground. Alternatively, you can use the 3:4:5 triangle method.

A BUILDER'S SQUARE — a large square made from metal — can be purchased from most major hardware stores. Because of the scale of most landscaping jobs, however, you will sometimes need to use one with a timber or metal straightedge which acts as an extender.

TIMBER SQUARE If you have a lot of setting out to do, you may find it easier to construct a simple timber square. Build it as large as is practicable, and ensure that it is as accurate as you can make it.

3:4:5 TRIANGLE METHOD

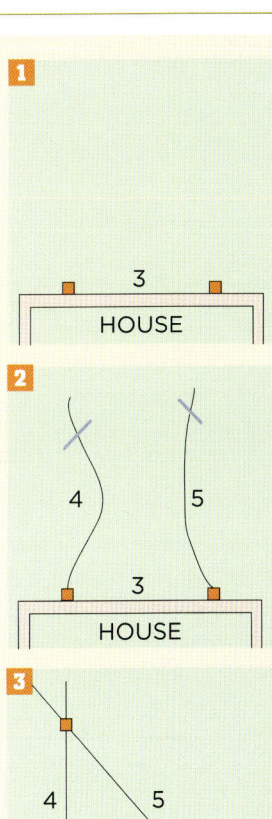

This technique relies on the fact that any triangle with sides in those proportions (e.g. 1.5:2:2.5 or 6:8:10) will be right-angled. Start by measuring and marking a 3-unit baseline — any units, metres or feet, will do — on the surface you are working to, such as the wall of your house, for example.

2 Attach string lines to each end of the baseline and make a mark at 4 units on one and at 5 units on the other (making sure you use the same units that you used in step 1, above).

3 Bring the two lengths of string together so that they cross where the marks coincide. This point will be at right angles to one end of the base. To make the triangle as accurate as possible, don't stretch the lines.

CIRCLES AND CURVES

Landscaping doesn't only involve straight lines, and two simple tools will help you to set out circles and curves.

TO DRAW A CIRCLE of any diameter on the ground, first hammer in a peg where you want the centre to be. Now attach a length of string, or the end of a tape measure, to a nail on top of the peg and scratch out the circle with a stick, or mark it with paint.

MARK CURVES with a length of flexible electrical conduit. The long lengths you can buy from electrical suppliers are ideal for marking out smooth curves for a path or garden bed. Position pegs at intervals along the conduit when you are shaping complex curves.

STAYING VERTICAL

Two basic tools are used to ensure that structures are vertical — a spirit level (page 77) and a plumb-bob.

A PLUMB-BOB is cheap and simple to use. You can buy a special weight with a pointed end for accurate work, but for everyday jobs any heavy weight on the end of a length of string will do.

DOING THE HARD GRAFT

Establishing levels

Once the rough set-out is complete (see pages 74–75) you are ready to establish levels — to work out how high things are relative to your existing house.

Imagine you were planning to install a 5m-wide wide area of paving off the back of your house, which is to end against a narrow garden bed and a brick retaining wall. Since the retaining wall is to be built first, you need to know where the top of its footing will be so that you can be sure that it will be concealed below the surface level of the paving when it is laid.

To do this, you would first of all work out what height you want the paving to finish against the back of your house. You would then extend a line out at this height (allowing for a minimum 1:100 fall to assist with drainage) to where the paving will end. This will now show you where the top of the paving will be against the retaining wall. You can now measure down from this point to where the top of the retaining wall footings will have to be. Once you know that, you can dig a trench to the correct depth and pour concrete for the footings.

There are a number of ways of establishing levels, and the various tools you can use are shown below and right. Basic instructions on how to use the equipment are given with illustrations on page 77, opposite. With the most accurate tools — the laser and auto levels — you may need some expert instruction. If you are hiring, ask for help at the store when you pick the tool up.

1. Auto level 2. Laser level and staff
3. String line 4. Tape measure 5. Spirit level

A SPIRIT LEVEL is the basic tool for levelling. Used with a straightedge, it will allow you to extend a horizontal line over distances of several metres, provided you work carefully. Always position your eye exactly over the bubble when centring it. Try not to bang the level, otherwise it may become inaccurate. To test for accuracy, place the level on a flat surface and check where the bubble rests. Now rotate the tool 180° and check again. The bubble should be in exactly the same place on both occasions.

A WATER LEVEL consists of a length of transparent hose filled with water — a little food dye added to the water makes it easier to see. This tool works because the surface of a body of water at rest is horizontal, and you can use this fact to transfer a height from one end of the hose to the other. Place one end of the hose so that the water is level with the height you want to transfer, and get an assistant to hold the other end of the hose at the place where you want the height transferred to.

A STRING LINE LEVEL is a small spirit level that can be suspended from a string line stretched between two points to check whether it is horizontal or not. Keep the line as taut as possible to minimise sag, and place the level as close as possible to the centre of the string line for the same reason. However, because of the inevitable problems with a sagging line, this device is not as accurate as other methods of levelling.

A LASER LEVEL is mounted on a tripod and spins around, sending out a harmless, perfectly horizontal, 360° beam. Where it strikes an object in its path, the beam shows up as a fine red line. Sometimes the line can be hard to see, in which case you can use an electronic receiver which is able to detect it. Laser levels are expensive, but they are easy to set up and use and you can hire them.

AUTO LEVEL You usually need two people to operate this high-tech level, which must be mounted on a tripod. One person looks through a small telescope on the level at a calibrated staff held by an assistant. By taking a series of measurements, you can work out the exact difference in height between various points around the garden. As with laser levels, auto levels are expensive to buy, but they can be hired.

DOING THE HARD GRAFT

Putting it all into practice

The easiest way to understand how to go about setting out a plan is to follow through in more detail the simple landscaping job described on page 76. As you will remember, this involves laying a 5m-wide area of paving at the back of a house, which is to finish against a double-brick retaining wall. Between the paving and the wall there is to be a channel drain to collect run-off and a narrow garden bed. The land slopes gently up from the back door of the house, so some excavation will be involved.

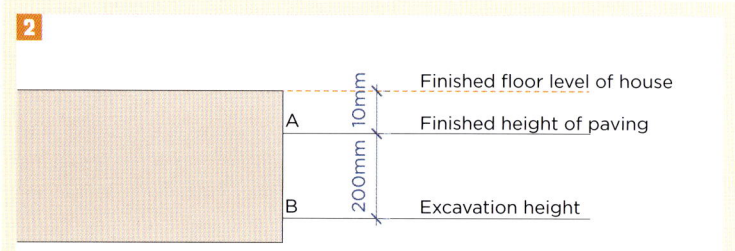

Establish and mark the finished height of paving, which for this job is going to be 10mm below the back step of the house. We will call this point A, and it will become a fixed datum, or reference point that you will continually refer back to. Now dig down 200mm to allow for the thickness of the paver (50mm), the thickness of sand bed (50mm), and the thickness of the roadbase or crushed rock sub-base (100mm). Mark this, which we will call point B.

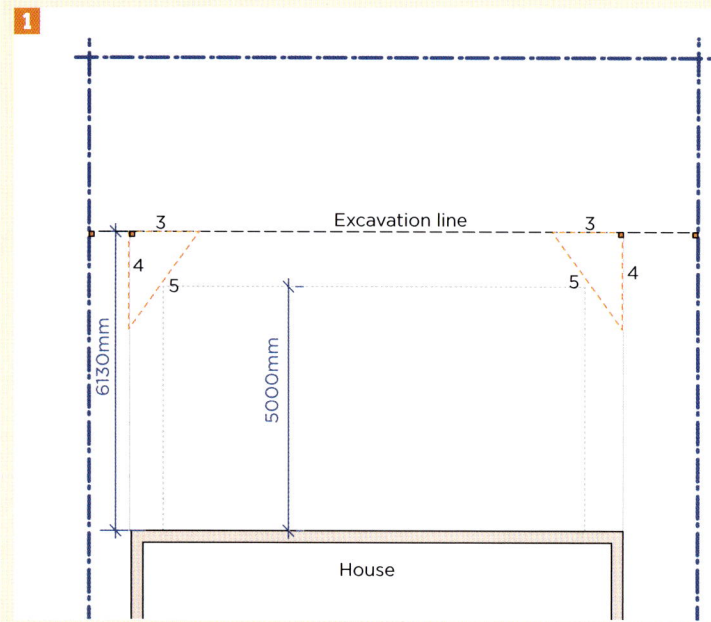

Using the 3:4:5 method described on page 75, measure from each side of the back of the house 6130mm to the excavation line and hammer in pegs. Alternatively, run string lines parallel to and along the side of the house walls. (This is based on 5000mm for the paving, 100mm for the channel drain, 300mm for the garden bed, 230mm for the wall, plus an allowance of 500mm behind the wall for free-draining backfill and comfortable working space during construction.) Run a string line between the pegs and extend the string out to the full width of the garden with two more pegs. Double check that your measurements are correct and spray along the string line with marking paint.

3

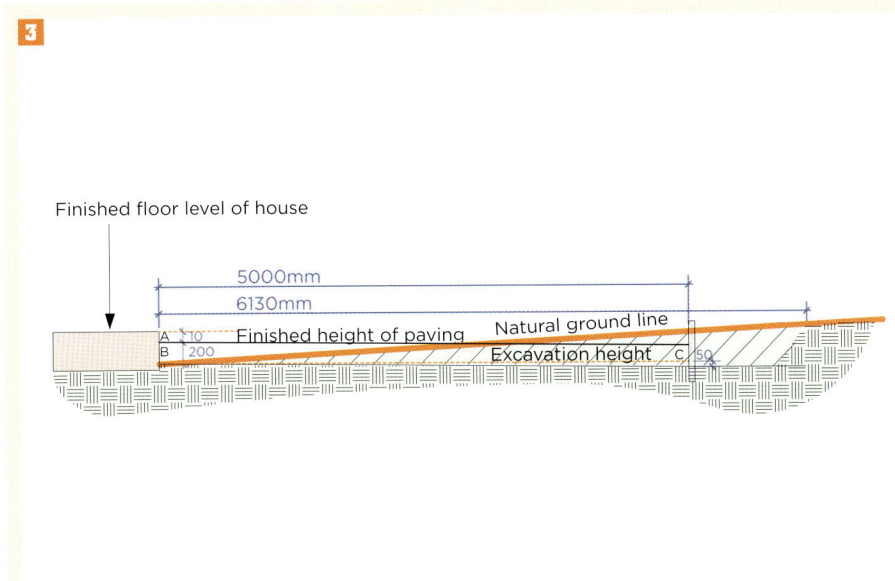

Now set a peg 5000mm from the house, where the far edge of the paving will be. Use your level to transfer the height of point B to this peg. Since the ground slopes up from the back of the house, you will need to dig out some soil to allow you to transfer the level. From the mark on the peg, measure down 50mm and mark point C. When you excavate to this point it will ensure that the paving slopes away from the house — has a 'fall' of 1:100 — so that water can drain away easily. Now remove the earth from the entire area at the back of the house, out as far as 6130mm.

Now set two string lines parallel with the back of the house, one of them 5215mm away and the other 5675mm away. These mark the two sides of the trench where the footing for the wall will be laid. The surface of the footing needs to be 75mm below the top of the paving, a point that will be 125mm above point C that you marked in stage 3 (above). Use your level to transfer the footing level to other pegs along the trench. The footing in this case will be 300mm deep, so dig down and install formwork ready to receive the reinforcing steel and concrete. The set-out is now complete and building work can get underway.

4

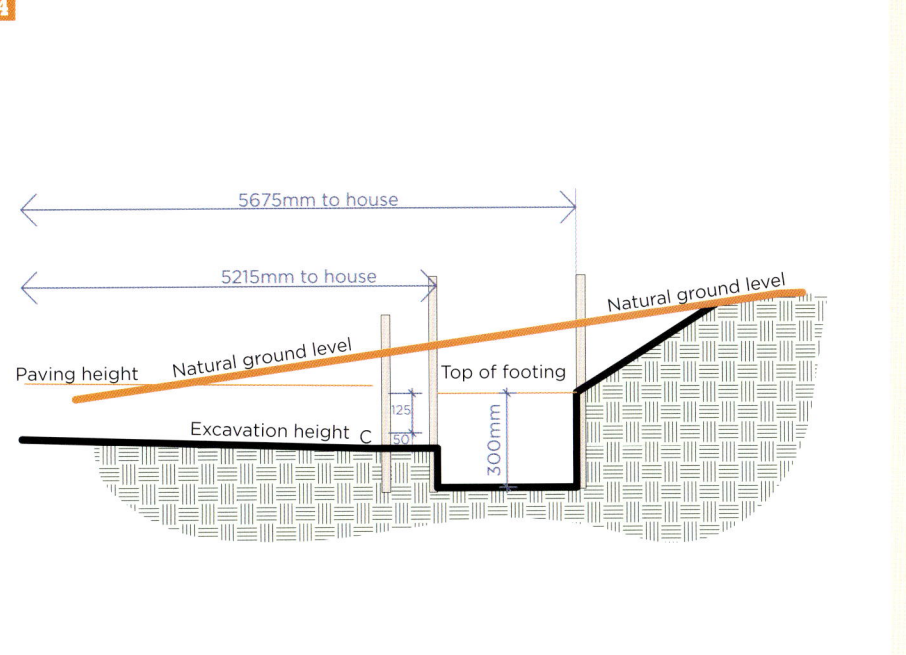

DOING THE HARD GRAFT

EXCAVATIONS AND EARTHWORKS

Time to tone those muscles

Most landscaping work calls for lots of digging — excavating for footings or paving, digging trenches for drainage, making major level changes and creating garden beds all involve moving quantities of soil from one place to another. This is a part of the job I really love. Call me mad, but I find something very satisfying about digging out loads of soil and watching the garden slowly being transformed as I do so. It can be hard work, but it can also be very rewarding.

The first thing to consider, therefore, is whether you are going to attempt to do the job by hand, or whether you're going to hire a machine. A major factor in making this decision is access. How are you going to get machines and materials into and out of the site? If there's no direct route into the garden, then it may be possible to negotiate access through a neighbour's property, perhaps by removing one or more fence panels. You will have to make good any damage after the job is finished, of course, but this may be cheaper than the alternatives. If direct access is impossible, then you might have to consider a conveyor belt or mechanical lifting system of some sort. Consult your local hire service.

If you do decide to hire a machine of any size — such as a bobcat; mini, medium or large excavator; backhoe or mini-loader — then I strongly recommend that you employ an operator as well. If you try to operate a bobcat by yourself, without any previous experience, you'll not only be slow, but you'll make a fool of yourself and there's a good chance you'll damage people or property as well. In most states you also need a licence.

Getting rid of excavated soil can be a problem. Try to retain as much as possible, but if you have to dispose of some, the easiest way is to bring in large skip bins or trucks that will take the waste away to a tip or landfill site. However, tip fees can be high, and this may be an area where you can save money. Bear in mind that while you want to get rid of some soil, someone else may be looking for some to use as fill. Advertise in local papers and talk to contractors, and with luck you might find a free tipping site (check that they have council consent); then your only expense will be transport costs. If that doesn't work out, it's probably easiest to get rid of soil in bins. Before you do so, however, consider saving any good-quality topsoil for use later on, provided you have room to store it. Good soil is expensive to buy.

In order to make plans for soil disposal, you'll need to be able to work out how much is involved. To pave an 8 x 5m area, for example, the amount of soil you will need to remove can be calculated as follows. The total area is 8 x 5m, which equals 40 square metres (m^2). Assume you need to dig down 180mm (0.180m) — to allow for 100mm of sub grade, plus 30mm of bedding sand, plus 50mm for the pavers — which means that the total volume involved is $40m^2$ x 0.18m, which equals 7.2 cubic metres (m^3). A large wheelbarrow holds around $0.1m^3$, so that's a total of 72 barrow loads.

Now you need to take into account one final factor to calculate the volume of waste accurately — what is known as the 'bulking factor'. This is caused by air pockets formed in the waste as you dig it out. Sandy soils bulk up very little, but clay soils can have a bulking factor of up to 30 per cent. The $7.2m^3$ calculated in the paving example above could therefore become nearly $9.4m^3$ in clay soil — another 22 barrow loads, and time to call up the bobcat operator!

The example above involved a simple rectangular shape, which was easy to calculate. To work out the volumes of more complicated shapes, break them down into a series of simple shapes — squares, rectangles and triangles — first to make the calculations easier.

Don't be nervous about hiring machines if they will make your life easier. A bobcat *(top and bottom right)* or mini-excavator *(bottom left)*, with an operator, is not that expensive to hire.

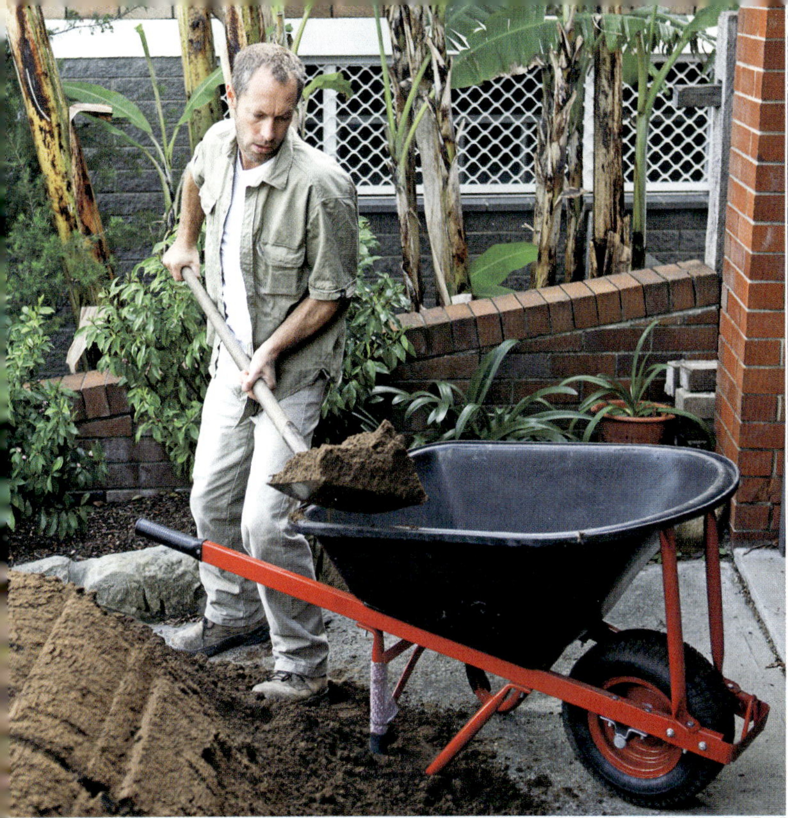

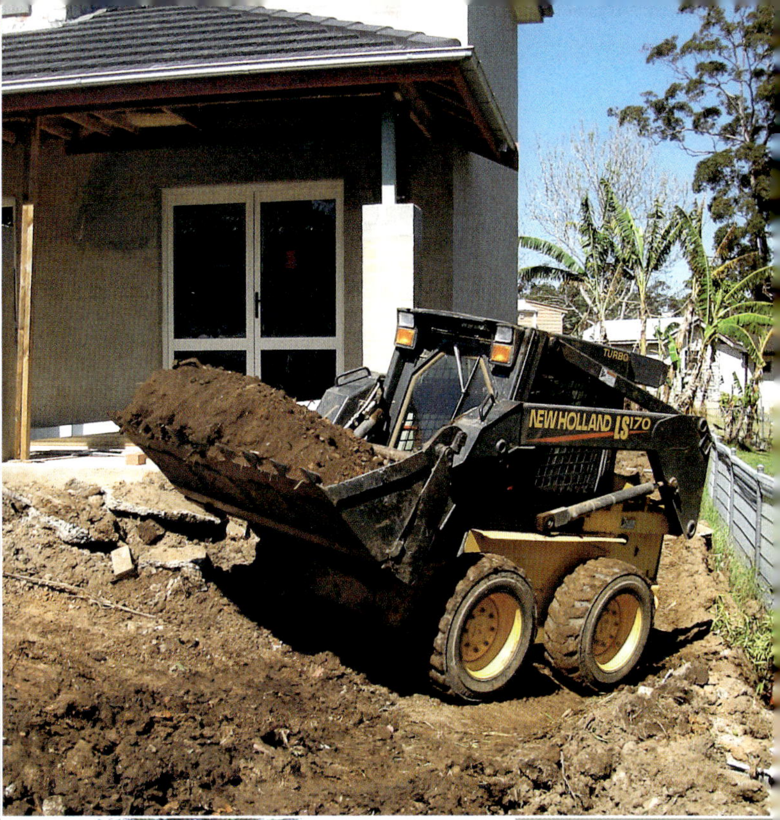

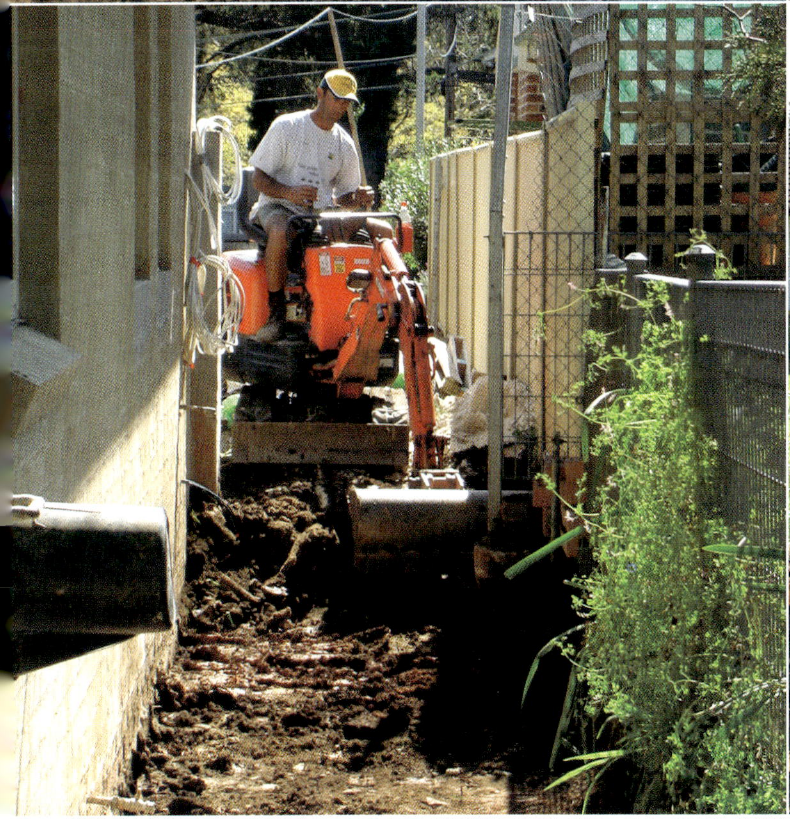

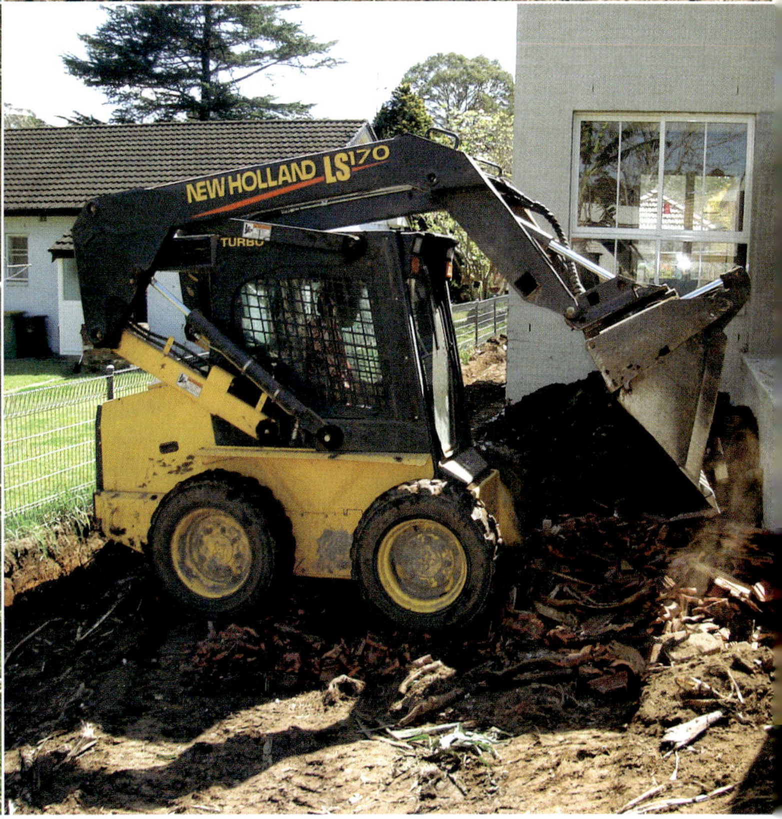

DOING THE HARD GRAFT

DRAINAGE

Managing water in your garden

Water plays a major role in any garden, and making the best use of it — including draining away any excess — is vital to the success of your landscaping plan. Factors you have to take into account when planning drainage include how porous your soil is, the shape of your land and also where the water is coming from. Rain is the obvious source, but you must also consider run-off from roofs, neighbouring properties and hard surfaces such as roads and paved areas. Water will naturally find its own path, and you have to work with that, especially if you change the topography of your garden by building structures such as retaining walls.

There are two aspects of drainage that you have to consider — surface and sub-surface. Surface drainage deals with water that runs off impervious surfaces such as paths and drives. That is why, when installing any hard surface, its fall or slope is important, since it will dictate where the water that drains from it will end up. Sub-surface drainage deals with water moving through the ground. That is particularly important when building structures such as retaining walls, where water needs to be collected and taken away to avoid pressure building up behind the wall, possibly leading to its failure.

Garden drainage can often be directed into the existing stormwater systems. However, you must take care not to cause pollution, which generally means installing silt traps to gather sediment. In many areas, work that is connected to the stormwater system must be carried out by a tradesperson with an appropriate licence. Check before you proceed.

Water can be collected and drained in many ways, and all the most common methods are illustrated here and on the following pages. Choose the solution that works best in your particular circumstances. When drainage becomes a problem — in situations where it is necessary to install tanks and pumps to remove excess water, for example — it's time to call in the experts.

Types of sub-surface drains

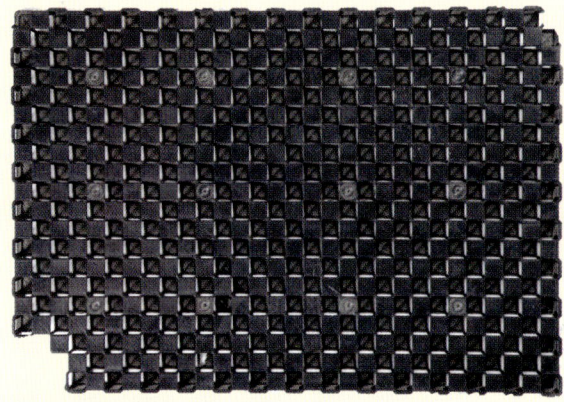

CELL DRAIN
Manufactured, modular cell-like sheets, boxes or strips.

USES Sub-surface water collection and storage, often used behind retaining walls and for draining roof-top gardens.

INSTALLATION Strip cell drains are installed in a similar fashion to agricultural drains (see page 85). However, these high-tech materials must be installed properly so read the manufacturer's instruction sheets before starting work.

SOAKAWAY PIT

A large hole dug in the ground, filled with loose rocks or old building rubble.

USES As the name suggests, water directed into a soakaway gradually disperses into the surrounding soil. Soakaways can be useful where no stormwater outlets are available, but they have limitations and can only handle so much water, especially when draining into a heavy, clay soil.

INSTALLATION Dig a large hole, install inlet pipes, line it with Geotextile fabric, fill with rocks or rubble and cover.

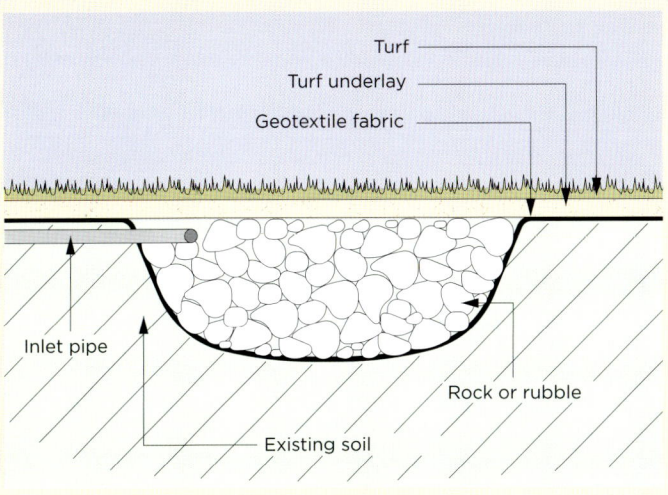

TRENCH DRAIN

A deep trench containing boxed cell systems, or trench covers, that creates an underground void where water can be held while it slowly disperses into the surrounding soil.

USES Good for use beneath lawns, especially where no stormwater outlets are available to connect to.

INSTALLATION Dig a trench to take the cell boxes or trench covers, deep enough to allow for a top cover of soil around 300mm deep. Line the trench with Geotextile fabric, install cells or covers, backfill with aggregate or blue metal, cover with more Geotextile and backfill with soil. Always read and follow manufacturer's installation instructions, if supplied.

Nige says...
DON'T TRY THIS AT HOME

You meet some funny people when you work in the landscaping business, which is one of the reasons why I think the job is such fun. I particularly remember one bloke many years ago who had made a bit of money and was determined to have a house and garden that reflected his status. I got the impression he was a bit of character on the first morning we started work. As we drove up his brick-red stencilcrete drive, with its flanking walls of bright yellow retaining blocks, he came out of the house to greet us. 'Tea or coffee?' he offered, 'or maybe you'd like something a little bit stronger?' Not many clients offer you alcohol at seven in the morning, especially when you're about to start working on their garden!

One of the jobs we had to do that week was to build a retaining wall with a drainage system behind it consisting of an agricultural pipe surrounded by blue metal. A suitable drainage point looked like a bit of a problem, but the owner insisted that he would take care of that, which was fine by me. It came as a bit of a surprise when we returned after the weekend to see him putting the finishing touches to the agricultural drain running straight onto a neighbour's property. 'You can't do that,' I told him, 'it'll flood their paving.' 'Don't be silly,' he said, 'I'm doing them a favour by watering their garden.'

DOING THE HARD GRAFT

Types of surface drains

DISH DRAIN
Open, U-shaped drain, constructed from pre-cast sections or formed *in situ* from concrete or some similar material.

USES Where surface water needs to be contained and directed, such as at the base of a wall where it meets a path or drive, or at the side of a path.

INSTALLATION Fix pre-cast sections into place, with concrete, ensuring a consistent fall in the desired direction, perhaps to a stormwater outlet. If you are forming the drain *in situ* from concrete, decide how wide and deep it needs to be to carry away the expected volume of water. Then use a length of pipe or a piece of plywood cut to shape to help form the dish.

PIT DRAIN
Manufactured, box-shaped pit with a grated cover; available in various sizes.

USES Collects surface water from hard surfaces and from lawns; good for changing the level and direction of a run of PVC piping or a channel drain; also acts as a silt trap.

INSTALLATION Once positioned in a trench, a pit drain can either be concreted into place or surrounded by gravel (river sand in a lawn). If setting in concrete, be sure to pre-drill for connecting pipes before pouring, because it will be impossible to do so afterwards.

GRATED CHANNEL DRAIN
Manufactured units, similar to dish drains in profile; available in a range of different sizes and materials, including heavy-duty polyurethane and stainless steel.

USES Excellent for draining the water that runs off from hard surfaces, such as paths or paving.

INSTALLATION Channel drains come in 1–3m joinable lengths, with corner sections also generally available. Cut the lengths as required, join them (sealing the joints with silicone) and set the completed drain in concrete or mortar just below the finished height of the adjoining surface, and hard up against it. If connecting to a stormwater drain, drill a hole of the correct diameter with a hole saw and seal the junction with silicone. If used with paving, mortar any pavers adjacent to the drain into place. Follow manufacturer's instructions, if supplied.

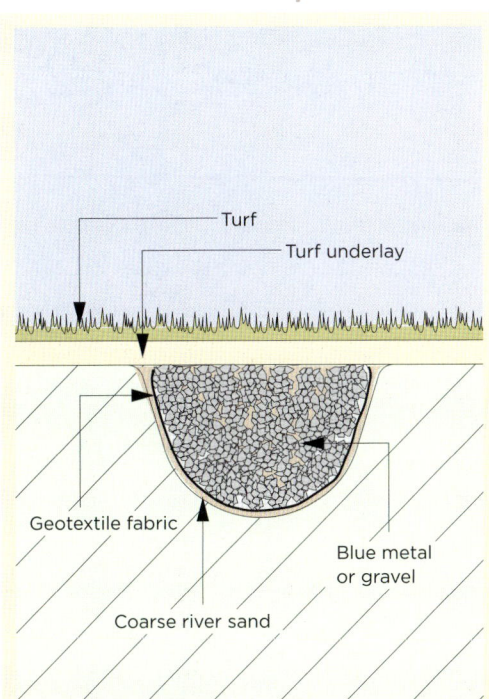

RUBBLE DRAIN

U-shaped trench filled with a drainage aggregate, such as blue metal.

USES Picks up surface water and some sub-surface water and helps to direct it; can be any width or length, either straight or curved; not practical in very sandy soils because the trench will collapse.

INSTALLATION Dig a trench — its size will vary according to the volume of water the drain is expected to carry, but about 300–400mm wide and deep is typical — and fill it with drainage aggregate, gravel or blue metal. The surface can be covered with turf or mulch. Rubble drains have a tendency to clog up with silt, and often need re-installing after a period of time; lining with Geotextile fabric will help to reduce silt build-up.

AGRICULTURAL DRAIN

Consists of a trench containing agricultural ('ag') pipe — which has holes in it — surrounded by river sand, gravel, blue metal or recycled aggregate. The drain sits below the surface, and can be covered with soil, turf or mulch.

USES A more advanced version of the rubble drain (see left), an agricultural drain also picks up surface and sub-surface water. Multiple agricultural drains can be installed, and they can also be configured in patterns — herringbone, for example — when used to drain areas such as lawns.

INSTALLATION Dig a trench 300–400mm wide and about 400mm deep. Agricultural pipe comes in two sizes — 65mm and 100mm — and the trench size varies accordingly. Maintain a constant slope, avoiding any undulations that might inhibit the smooth flow of water. Line the bottom of the trench with 50–100mm of river sand (my own preference) or alternatively 10 mm of gravel or blue metal. Lay the pipe in the middle of the trench, keeping it level and straight, and cover it with more river sand, gravel or blue metal to a depth of about 100mm, then backfill with soil. In some cases, Geotextile fabric is used to line the entire trench, or just its base.

DOING THE HARD GRAFT

FOOTINGS AND FOUNDATIONS

Providing the necessary support

The terms footings and foundations are often confused, but they mean quite different things. Footings, usually made from concrete, brick or stone, support walls, piers and other structures. They distribute the weight of the wall or structure above and therefore need to be strong enough for the particular job in hand.

Foundations are the natural ground materials that support the footings and the structures that are built on them. Good foundations are soils with a strong structure that are not influenced by changes in moisture content and temperature. Sandy soils and rock make good foundations. Unsuitable foundation material — such as reactive clay soil, fill, or soils high in organic matter — can lead to movement, and may even cause the structures built on them to fail. Where possible, remove all unsuitable foundation material before installing footings. This may involve digging a trench deeper than necessary and then filling it with sand or another good foundation material to bring it back up to the correct depth. A natural stone outcrop will make a good foundation, provided it is bedrock and not just an isolated, buried boulder that may move.

Although the focus in landscaping is mainly on smaller structures, such as garden walls, retaining walls, barbecues, planter boxes and water features, the footings you build still have to be strong enough for the job. They need to be consistent in thickness, level, and are always wider than the wall built on them — about double the width is generally recommended.

If you have any doubts at all about the strength and suitability of the footings you are planning for a particular job, consult a qualified engineer. Some councils require engineer's plans and certificates for a wide range of structures as a matter of course. It is particularly important to have an engineer design any major structure where failure or collapse could have serious implications for either yourself and other occupants of your property or for neighbours. If you do have an engineer design and approve a structure then it will at least ensure your peace of mind, since then any errors or failures will become their responsibility. The strength and suitability of major garden structures may also become an issue when you try to sell your property, so it pays to always err on the side of caution.

BRICK AND STONE FOOTINGS

Most footings are made from concrete, but other materials can sometimes be used. For low walls and edging, for example, a simple brick or stone footing may be adequate. This consists of a few layers of bricks or stones laid below ground level, with the footing wider at its base and gradually becoming narrower. This is often a good way of using up a supply of old bricks recycled from demolition work on a site.

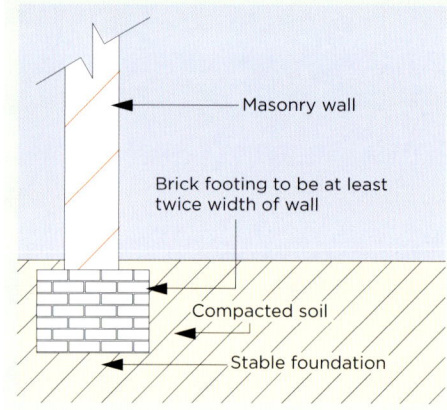

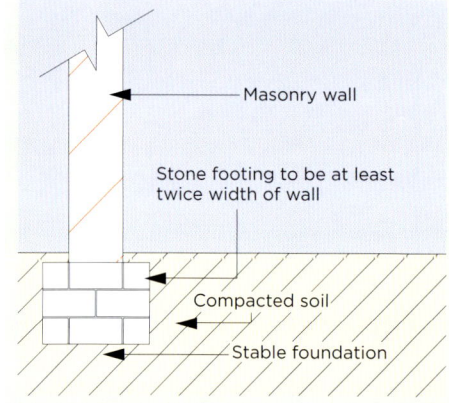

CONCRETE FOOTINGS

Concrete can be used on its own, but it is much stronger when used with reinforcing mesh. I strongly recommend that you use mesh every time, unless an engineer tells you to do otherwise. The types of concrete footings most used for landscaping jobs are strip, raft and blob. For information on buying and using concrete, as well as instructions for preparing and laying simple footings see pages 88–89.

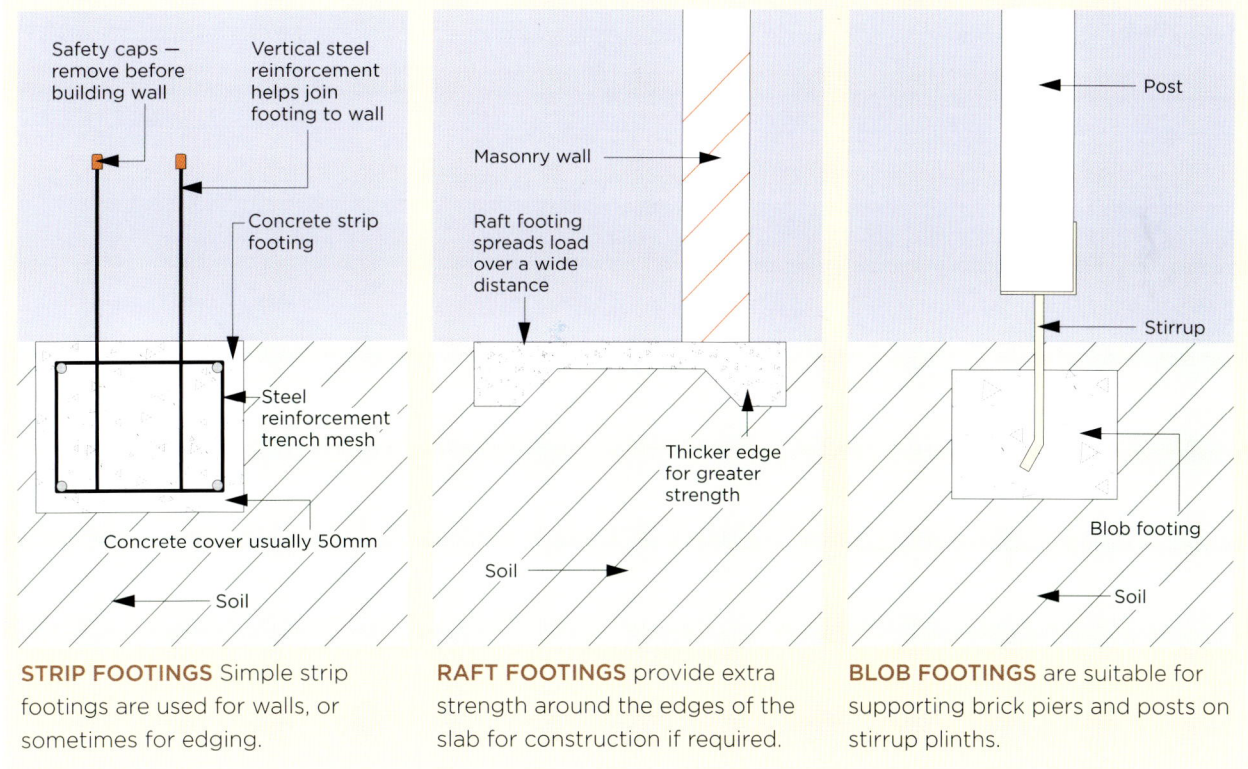

STRIP FOOTINGS Simple strip footings are used for walls, or sometimes for edging.

RAFT FOOTINGS provide extra strength around the edges of the slab for construction if required.

BLOB FOOTINGS are suitable for supporting brick piers and posts on stirrup plinths.

DOING THE HARD GRAFT

Building simple concrete footings

There's nothing better than watching the concrete truck drive up on a sunny morning knowing that you're ready to go, with the formwork securely in place, enough hands to help out and all the tools to hand. The opposite is a nightmare, and that's the point: you need to be organised when it comes to concrete.

There are two ways you can go about concreting: mix it yourself or get it delivered premixed. I would recommend premixed for all but the smallest jobs because it saves a lot of mucking around, and in some circumstances it is the only realistic option. There are also two ways in which you can go about placing concrete: you can barrow it, or you can pump it. A concrete pump — which delivers the concrete through rubber hoses to the job — is a must for sites where access is difficult. It adds a bit to the cost, but is much easier on your back.

WHAT IS CONCRETE? Concrete is the standard building material used for making garden paths, drives, slabs and footings. It is strong — especially when used with reinforcing steel — relatively cheap and easy to install (on small jobs). It is made from Portland cement, sand and aggregate (stones), combined in various proportions, according to the type of mix required for the job at hand. The strength of concrete is measured in megapascals (MPa), with a 20MPa mix a good standard for general garden jobs, such as paths and footings for low walls. For anything more complex, such as a patio slab, get advice from an engineer and consider calling in the experts. If ordering premixed concrete, tell the suppliers what you are doing and they will advise on the best mix for the job.

For small jobs, such as setting fence posts, you can buy the raw materials separately or bags of the appropriate dry mix. For dry mix, the exact amount of water you need to add will be given on the back of the bag. Don't use too much water, or you will weaken the concrete. If mixing the raw materials yourself, choose a mix appropriate for the job. For a general purpose mix with a strength of 20MPa, use (by volume) 1 part cement to 3 parts coarse aggregate and 2.7 parts coarse sand.

STEP-BY-STEP THROUGH A TYPICAL PROJECT

The footing for a small freestanding garden wall is the sort of concreting project that could be tackled by any reasonably handy amateur. For retaining or load-bearing walls, you will need to get advice from an engineer on the proper dimensions. Some councils will insist on engineer's drawings anyway.

Concrete is also used for other jobs around the garden, such as paths and drives. For details on how to prepare and lay a slab, see pages 106–107.

The size of the footing will depend on the size and type of wall, and the type of soil in your garden. As a rough rule, the width and depth of the footing should be twice the width of the wall that's going to be built on it. Therefore, the footing for a low double-brick wall, 230mm wide, should be about 460mm wide and 460mm deep. The length of the footing should equal the length of the wall plus its thickness.

There are four stages: setting out and excavating, installing the formwork, installing the reinforcing steel, and pouring and finishing the concrete.

PG *recommends*

ORDERING CONCRETE

Concrete is ordered by the cubic metre (m^3) and comes in $0.2m^3$ increments. So, if you've calculated that you need $1.5m^3$ for your footing (length x width x depth) then you need to order $1.6m^3$. There's nothing worse than running out of concrete, so it's always better to order a bit too much rather than too little.

SETTING OUT AND EXCAVATING

The basic tools and techniques for setting out a job are given on pages 74–79. Take your time over this stage of the job as a mistake can be costly and time consuming to rectify once you start digging. The actual excavations can either be done by hand or with the help of a machine (see pages 80–81). If you do decide to hire a machine and operator, check first to ensure that there is enough space to work in, and also that your arrangements for the storage and disposal of soil are satisfactory.

INSTALLING THE FORMWORK

The formwork — made from timber boards or plywood nailed to wooden pegs — supports the concrete while it hardens. It needs to be strong enough to take the weight of the concrete — which weighs 2.24 tonnes per m^3 — so don't skimp on pegs. Plywood is good for creating curved shapes, but it needs plenty of support because it is not very thick. Set the top of the formwork to the exact finished height, since this enables you to use the boards for screeding.

INSTALLING REINFORCING STEEL

The steel mesh must be positioned correctly and firmly supported. Plastic support chairs keep the mesh off the ground. Position them no more than 600mm apart in every direction and attach them to the steel with tie wire. Make sure that adjoining sheets of reinforcing steel overlap each other by two grid squares, and also that the steel is at least 50mm away from the formwork to ensure it's completely covered by concrete. If the steel is exposed, it will rust and crack the concrete.

POURING AND FINISHING

Get the concrete into the job and ensure good placement by settling it with a shovel or float to avoid air pockets or voids. Then level it roughly with a shovel and finally screed it with a straightedge placed on top of the formwork. Leave the concrete to cure for seven days, unless you want to finish it further, which is only necessary when it can be seen (pages 106–107). Don't play with the concrete. The more you move it, the more water will come to the surface, weakening the top layer.

DOING THE HARD GRAFT

GARDEN STEPS

Proportions that work best

Steps have a great unifying effect in a landscape. In addition to serving the practical purpose of allowing access between different levels in a garden, they also provide a focal point to highlight aspects of a design. In my own garden I have a set of curved stairs that are often used for informal seating by family and visitors, even when garden furniture is readily available.

In practical terms, steps need to be placed where they will be used. I have seen many poorly planned gardens where the steps are ignored, with people taking a short cut through garden beds instead. Steps — and paths for that matter — need to be placed where people will walk. We're a lazy lot as a rule, and have a natural inclination to take the shortest route between two points, along what are known to landscape planners as 'desire lines'. If you ignore desire lines when you prepare a plan, then your turf or garden beds will end up with a goat track running through them.

There are many styles and forms of steps, and they can be constructed out of a wide variety of materials. I will look at a couple of simple construction techniques for small flights of steps (pages 92–93) that any reasonably practical person should be able to build. For anything more advanced, such as a flight of timber stairs, I would recommend that you call in a professional builder. Poorly constructed steps and stairs can be dangerous and may cause accidents, which could have serious consequences for a garden owner.

Centuries of experience in step building have shown that certain proportions work best, and these have been enshrined in the building code that applies to steps. It states that all the steps in a flight must have the same riser (vertical height) and the same tread (horizontal measurement). The rule of thumb that applies is simply twice the riser plus the tread should equal 625–675mm, or 2 × riser + tread = 625–675. This is known as the riser–tread ratio, and the easiest way to apply it is to calculate the total difference in height between the two areas to be linked by the steps and divide this figure by 150mm, which is a standard height for a riser. This will give you the number of risers needed to reach the top. Now divide the number of risers into the original height to give the actual riser height.

A PRACTICAL EXAMPLE Say you wanted to build some steps from a patio down to a lawn, where the height difference is 480mm:
- The total height is 480mm.
- Divide 480 by 150, which equals 3.2.
- Round this out to the nearest whole number = 3 risers.
- Divide 480 by 3 to give the actual height of each riser = 160mm.

Now you can use the riser–tread ratio we discussed above (2 × riser + tread = 625–675) to work out the depth of each tread.
- The risers in our example are 160mm, so 2 × 160 = 320mm.
- Subtract 320 from the minimum distance, 625 − 320 = 305mm.
- Subtract 320 from the maximum distance, 675 − 320 = 355mm.

Therefore the treads must be between 305mm and 355mm wide to comply with the rule. In making a final choice, bear in mind that treads should not be less than 300mm wide, because it is difficult to place your foot on a tread any narrower than that.

You can now calculate the length of the flight of steps (or its 'going') by adding the width of the treads together. Our example has three risers, so it will only need two treads (a total of 610–710mm), although if the height change is over a greater distance you can always add an extra tread at the top of the flight.

LANDING Firm, flat area at the top and bottom of a set of steps or stairs.

GOING The 'going' of a set of steps is the total horizonal distance between its top and its bottom.

RISER The vertical portion between one step and the next above or below it.

TREAD The horizonal portion of a step, on which you place your foot when climbing.

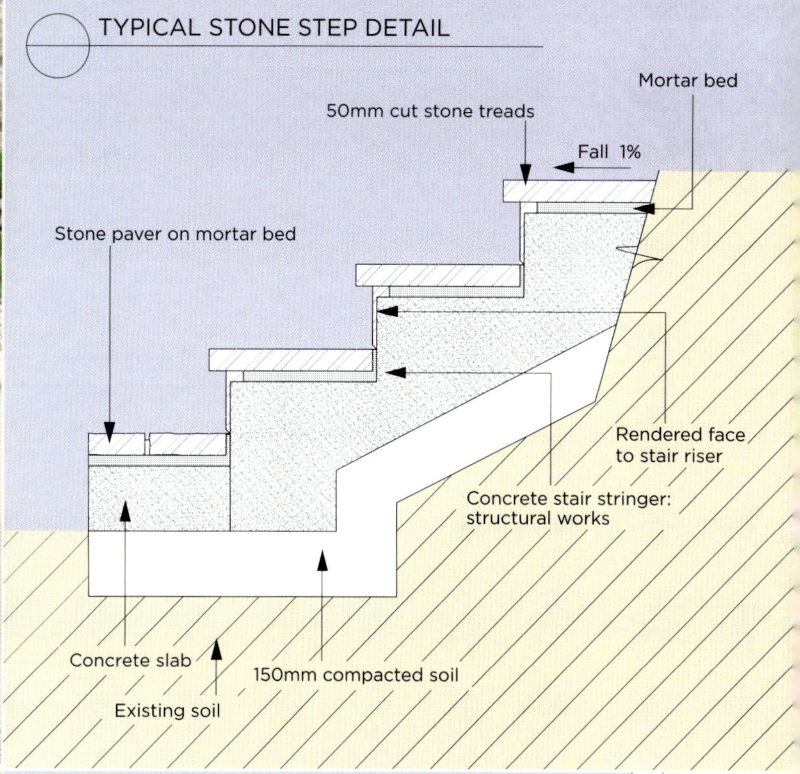

TYPICAL STONE STEP DETAIL

- Mortar bed
- 50mm cut stone treads
- Fall 1%
- Stone paver on mortar bed
- Rendered face to stair riser
- Concrete stair stringer: structural works
- Concrete slab
- 150mm compacted soil
- Existing soil

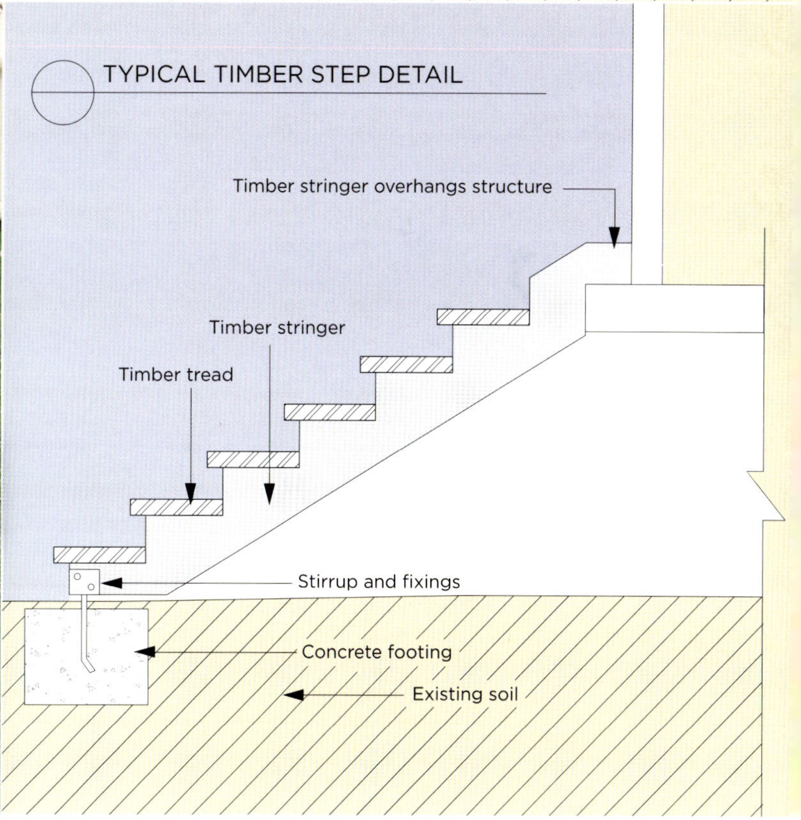

TYPICAL TIMBER STEP DETAIL

- Timber stringer overhangs structure
- Timber stringer
- Timber tread
- Stirrup and fixings
- Concrete footing
- Existing soil

DOING THE HARD GRAFT

Building simple stone block steps

Quarried stone is readily available in most areas of Australia — in Sydney the local product is sandstone, in Perth it is limestone and in Melbourne bluestone. These materials have been used in building projects for generations, and because of this it is often possible to buy recycled stone for use in construction jobs in your garden. Recycled materials bring a sense of age to a garden, introducing an interesting new dimension to the landscape. Therefore, if you are planning to use stone to build some steps, it may be worthwhile visiting local recycling centres or demolition yards to see if you can pick up something suitable.

The first job is to calculate the tread and riser dimensions for the steps you are going to build (pages 90–91). Once you have decided on that, and marked out the site (pages 74–79), you are ready to set the first stone into place to form the bottom step. If the stone blocks you have are higher than the planned riser height, you will have to excavate to lower the blocks to the desired level. Generally, if large blocks are being used, you can lay them directly onto the existing soil. However, if you don't think the soil is stable enough to provide a firm support, dig out to a greater depth and fill the area with crushed rock to form a solid base. With the first step in place, backfill behind it with roadbase or crushed rock, ready to take the next block. Continue in this way until the steps are complete.

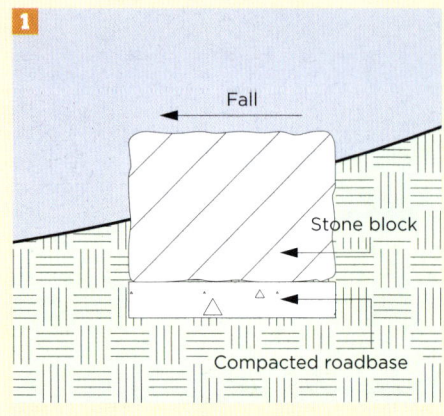

Excavate as required and position the first block, allowing a fall to the front.

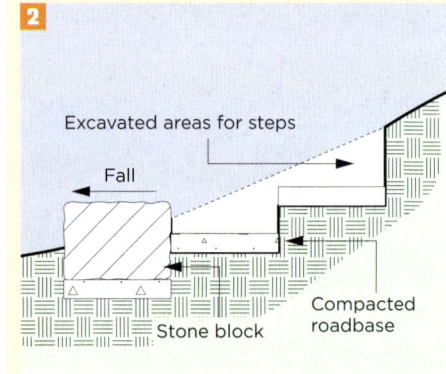

Prepare the area behind the first step, ready to take the second step.

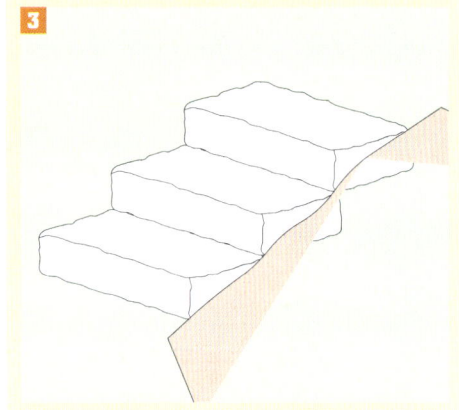

With all the steps in place, shape soil neatly around the step ends.

Building simple sleeper steps

Standard 150 x 75mm treated pine sleepers lend themselves to step construction, both because of their durability and because of their standard size. Laid on its edge, a sleeper forms a riser that exactly matches the recommended standard height of 150mm, while two sleepers side by side make a comfortable tread of 300mm. When determining the width of a set of steps to be made from treated pine, remember that in most cases sleepers are sold in standard lengths of 2.4 or 3m, so you can minimise waste by thinking ahead.

The easiest way to make sleeper steps is to prefabricate each step as a separate unit, made up from one sleeper of the desired length on its edge to act as a riser, two sleepers of the same length, side by side behind the riser to act as the tread, and two 300mm-long side pieces to act as stays. Screw all the pieces together with bugle head screws, and bevel the edges of the timber to reduce the risk of splinters. On very wide steps, add a centre stay, in addition to those at the sides, to stop the timbers from twisting.

Mark out the site (pages 74–79) and excavate ready for the bottom step. Place the prefabricated step in place and level it carefully, allowing for a slight fall to the front to stop water from pooling. Then simply stack subsequent steps behind the first, with the front of each riser fixed with screws or nails onto the back of the tread in front of it. Backfill with gravel to allow water to move away from the timber, preventing rot.

Sawdust from treated pine should not be inhaled, so always remember to wear a mask when cutting it. Apply a sealing agent to all cut surfaces to restore the timber's resistance to decay.

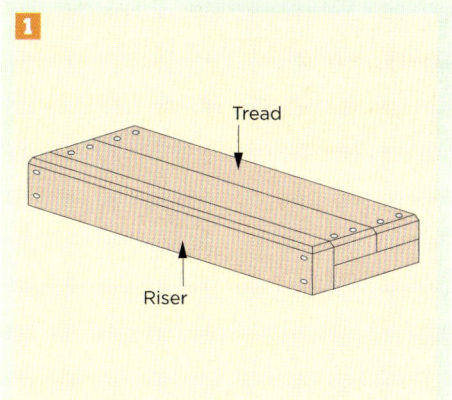

Prefabricate each step, screwing the lengths of timber together.

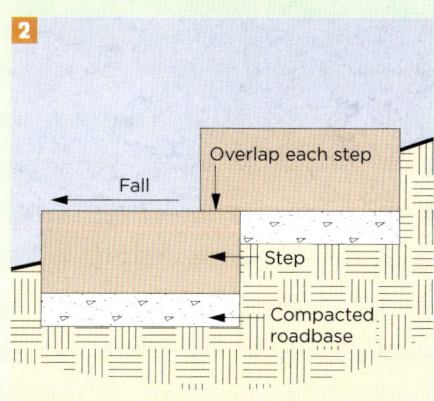

Position the first step, allowing a slight fall towards the front to shed water.

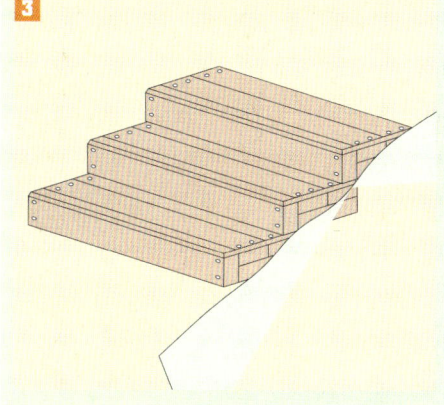

With all the steps in place, shape soil around the step ends for a neat finish.

DOING THE HARD GRAFT

WALLS AND FENCES

Versatility is the key

Walls and fences can have many different functions in a landscape: they can be used to define boundaries and mark off areas within a garden, as well as between the garden and the house; they can be used to hide eyesores or accentuate special features; they can be used to retain soil when levels change; and they can also be used for seating. Depending on use, they can be freestanding or incorporated into the landscape as a whole.

Materials and finishes vary enormously, so there should be no problem choosing something that harmonises with the overall effect you are trying to create. In the pages that follow, I show you how to build a simple retaining wall and a fence, but there are many other possibilities. Before tackling something really ambitious, however, pause to make an honest assessment of your level of skill as a builder. A poorly built wall or fence is just an annoying waste of time and money, but the failure of a badly designed retaining wall can have serious consequences for yourself and your neighbours.

Before building any wall or fence, make sure that you check with the local council. There may be regulations that govern the height and style of structures you can build in your area, and it's important to be aware of any restrictions before your plans are too far advanced.

There are four important points to bear in mind when planning any retaining wall — the 'angle of repose' of the material being retained, footings and foundations, backfilling and drainage.

The 'angle of repose' of a material is the natural slope it would eventually reach if you simply left it alone. Sand, as you might imagine, has a low angle of repose, while clay or shale have high angles of repose. When excavating, it is important to remove material to the angle of repose, otherwise soil will fall into your trenches.

Firm footing and foundations are always the most important aspects of any retaining wall. Inadequate preparation will not only look unsightly, but may lead to the wall failing. Make sure that the excavations are deep enough to accommodate the base of the wall, and that the bottom of the wall has solid ground in front of it to prevent it from slipping forward. Use concrete, brick or stone, or in some cases crushed rock or roadbase that has been thoroughly compacted.

When backfilling a wall, it is important that you use a free-draining material such as sand or gravel. The most common reason why walls fail is because of inadequate drainage. One cubic metre of water weighs 2.25 tonnes, so it is easy to imagine the enormous pressure that can build up behind a poorly drained wall during heavy rain. You can also assist drainage by ensuring that the joints between blocks that make up the wall are free — what is known as an 'open wall' — because it allows water to move easily through its face. It is also always a good idea to lay an agricultural drain at the back of the wall to collect water and move it to a convenient drainage point.

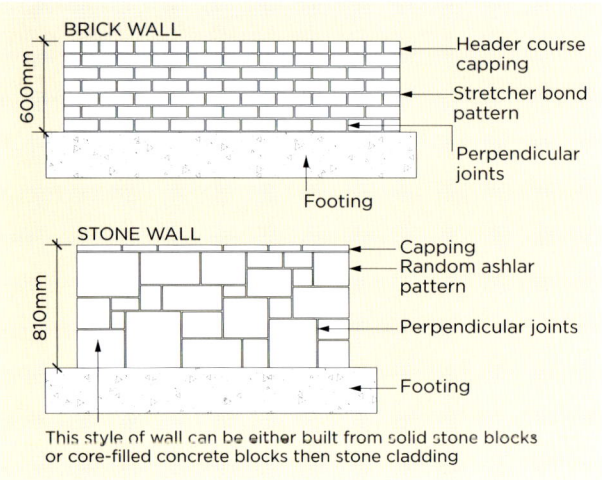

Walls and fences are usually among the most prominent and visible structures in a garden. It is therefore important that you choose construction styles and materials that complement other elements in the overall landscape design.

DOING THE HARD GRAFT

Building a simple retaining wall

Treated timber makes an attractive and serviceable retaining wall, and one that is fairly straightforward to build. There are many ways of using the timber, but the simplest is known as a post and whaling wall (whalings are the backing timbers which are laid horizontally behind vertical posts). Both round logs and sleepers work equally well with this type of construction.

The height of the wall — check with your local council to find out if they have imposed any restrictions on height — should not exceed 1.5m without the benefit of expert advice from an engineer. In most cases, a post approximately every 1200–1500mm will be sufficient. Since most treated pine products are sold in 2.4 and 3m lengths, this means that for each section of whalings there will be a post at either end and one in the middle.

MODULAR WALL SYSTEMS

Many of the large concrete brick and block manufacturers produce a range of modular blocks specially for building retaining walls. They are designed to lock together easily, and for beginners they are probably the simplest and quickest way of building a wall. While generally not as attractive as natural stone, modular blocks are cheaper to buy and easier to work with. Building instructions and some attractive project ideas are usually available on the block manufacturers' websites.

1 Lay out and mark the position of the wall, using the methods described in the sections on establishing levels and setting out (pages 74-79). Dig out the area behind the wall, taking care to work to the angle of repose for the material concerned (page 94), and also to avoid any existing pipes or sevices. Make sure you allow plenty of room behind the wall for the installation of drainage, and also to provide adequate room for working. There is nothing worse than a cramped work space, and I usually allow at least 300–500mm.

The finished area where the base of the wall will rest needs to be flat and level so that the first whalings will sit properly, but you won't generally need to dig out and fill with crushed rock or roadbase.

Excavate post holes (the width of the posts plus 200mm) at both ends of the wall, the same depth as the height of the wall. Put 100mm of blue metal in the base of the holes.

Centre a post in each hole, check that they are vertical, front and back and side to side, and backfill the hole with concrete. Provide support for the post while the concrete sets.

Run a string line between the end posts and mark, dig and concrete into place all the intermediate posts. Slope the surface of the concrete away from the timber.

Position the whalings behind the posts and nail or screw them into place. If you cut treated pine, remember to treat the cut surface to restore its resistance to rot.

Install an agricultural drain (page 85) behind the wall and connect it to pipes to carry away waste water. Backfill with a free-draining material such as sand or gravel.

DOING THE HARD GRAFT

MASONRY WALLS

Brick and concrete block walls are the most difficult structures for a home builder to make. Unless you have had some experience, I would suggest that you call in a professional to build a brick wall. Having said that, bricks are very adaptable, and there is no good reason why a skilled amateur, working carefully and methodically, should not be able to build simple brick structures. Here are a few pointers if you're feeling adventurous:

- Always keep the courses level — the easiest way is with a spirit level and water level (see page 77).
- Build the corners first, then fill the space between, using a stringline to keep the course level. Keep the joints consistent — they should be approximately 10mm wide. Check the distance between the corners and work out how many full bricks will fit into the available space. Avoid a single small cut by planning two larger ones instead.
- Try to match any existing brickwork on your property. The fewer different elements there are in a landscape, the better.
- Mix small amounts of mortar at any one time, and don't use mortar once it has begun to dry out.
- Keep the mortar moist and well worked — it should have the consistency of whipped cream.
- Before starting on the actual job, practice your technique until you have mastered the basics. For trial runs, use 1 part lime to 6 parts brickies sand. This mix behaves like normal mortar, but can be easily be pulled apart when you have finished.

Building a timber fence

Fences not only act as boundaries around a property, but can also be used within a garden for a range of other purposes, such as: helping to screen off unattractive domestic service areas, like washing lines or garbage bin storage; providing privacy; protecting potential hazards such as swimming pools; creating free-standing structures; or helping to draw attention to elements in a landscape. Timber or metal provide the superstructure for most fences, and this can be clad in many different materials, ranging from traditional palings, bamboo screens and natural brush to galvanized steel, chainlink wire and fibrous cement sheeting. What you choose will depend on what the house is made from, other elements in the garden and the overall 'look' you are trying to create.

The simplest type of fence to build — and perhaps the one most often seen in suburban Australia — is the timber post and rail construction with a cladding of timber palings. You can buy a fence kit of this type from a fencing supplier and it will come complete with cladding, posts and rails. Place posts at regular intervals, roughly 1.8m apart, and join rails at a post, staggering the joints so that top and bottom joins do not coincide. Depending on the weight of the fence and the type of soil in your area, you may not need to concrete the posts into place. If you do backfill with earth, compact it at regular intervals to ensure that the posts are set as firmly as possible into the ground.

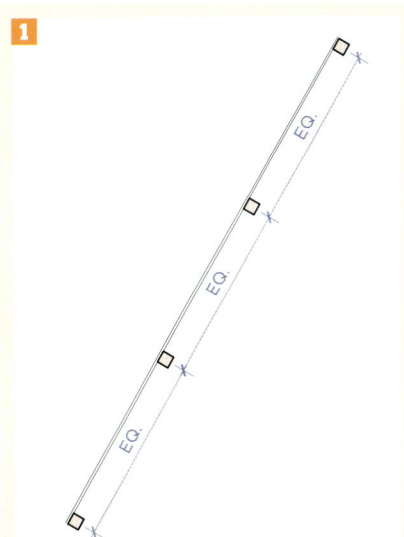

Lay out and mark the position of the fence, using the methods described in the sections on establishing levels and setting out (pages 74–79).

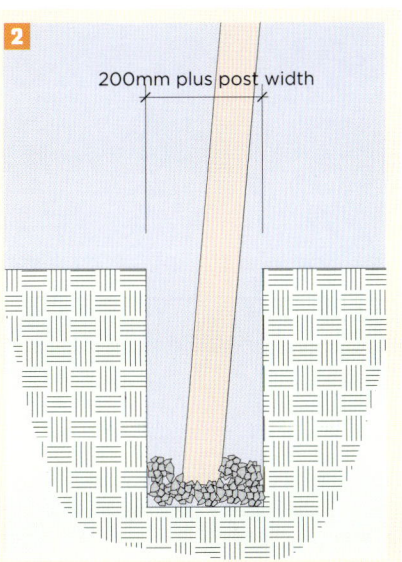

Excavate 600mm-deep post holes (the width of the posts plus 200mm) at both ends of the fence. Put 100mm of gravel in the base of the holes.

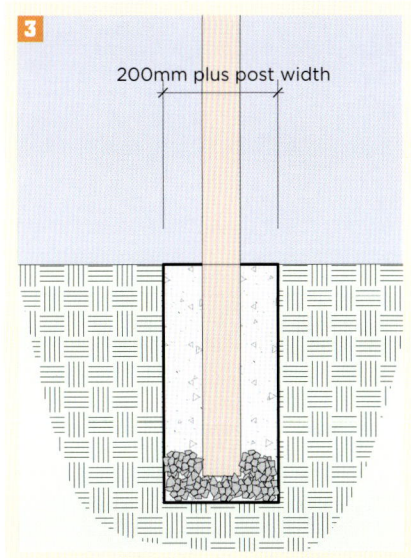

Centre posts in the holes, backfill with concrete or thoroughly compacted soil, depending on the type of soil and the weight of the fence.

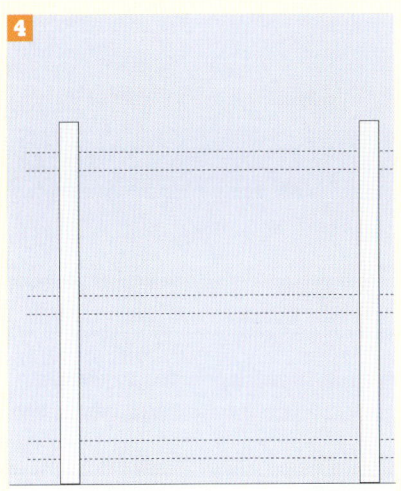

Run a string line between the two end posts and mark the positions of intermediate posts. Dig holes and fix all posts into place.

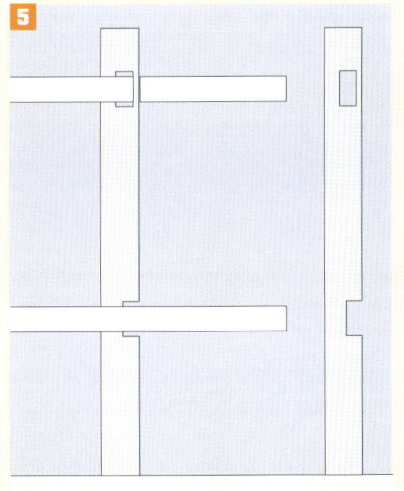

Prepare and fix the rails, either by passing them through pre-cut holes or by slotting them into recesses you have cut into the faces of the posts.

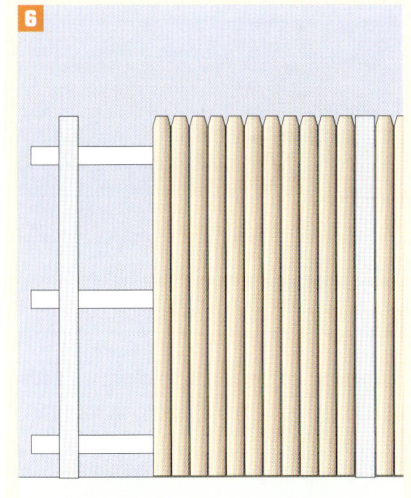

Apply your chosen cladding and finish as required. If you are using treated timber, remember to seal any freshly cut surfaces to prevent rot.

DOING THE HARD GRAFT

PATHS AND PAVING

Helping to bring it all together

Paving is undoubtedly one of the most important structural features of a landscape. Not only does it perform the obvious function of providing an all-weather surface for access and entertaining, but it also provides a unifying element within the garden — helping to link areas with strong accent lines, colour and form.

There are many different forms and styles of paving, and the first job is to select a type of paver to suit the style of garden you are trying to create. Small clay pavers work well in some garden styles, while large-format pavers look better in others. Have a look through magazines, visit some display homes and tile showrooms or walk around your local neighbourhood to help you decide which pavers will work best with your design.

Depending on the type of paver and the application it will be used for, there are a number of different laying techniques that are described in detail on the following pages. However, there are some basic tips that apply to most kinds of paving (see right).

STRETCHER By laying pavers in a stretcher pattern across a site, you can make the back seem closer. If you lay pavers straight down a site, the strong lines make side boundaries appear to be closer together.

HEAVY DUTY When using 230 x 115mm pavers for driveways, a herringbone pattern will produce a tightly interlocked surface that will be better able to cope with the impact of heavy vehicles.

THICKNESS Check with your supplier before making a purchase, pavers differ in thickness according to their intended use. Pavers for driveways must be thicker than those intended only for foot traffic.

WATER You must allow for drainage, since more than 96% of the water that falls onto an area of paving flows off it. You must collect this water and dispose of it properly, using drains where appropriate.

GAPS Follow maker's instructions when laying pavers. Most dry bed pavers must be laid with a 2–3mm gap between to allow for movement, especially on drives where too much movement can cause chipping.

LINES When laying pavers, set out a grid of string lines to help you in keeping the pattern perfectly square and to prevent lines from wandering. Crooked lines — even those that waver slightly — will look unsightly.

MAINTENANCE Some pavers may need to be sealed to stop staining. Check with your supplier to find out what maintenance is required for the product you are considering, before making a purchase.

DOING THE HARD GRAFT

Dry bed paving

Sometimes referred to as flexible paving, this is the system commonly used for laying brick sized pavers. The pavers are laid on a bed of gritty sand approximately 30–50mm thick, and are locked into place with fine sand brushed into the joints. The great advantage of this system is that the pavers can be lifted and replaced if necessary.

Foundations are very important when it comes to laying pavers. The material must be consistent, free from large amounts of organic material and non-reactive — it won't expand or contract excessively. If necessary, remove the material that is there and fill with roadbase or something similar to form a level surface. The area to receive the paving must be excavated to the thickness of the paver, plus the depth of the bedding sand and the sub-base.

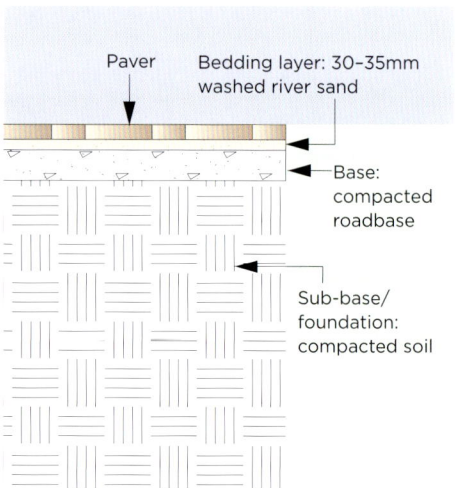

PREPARING THE SUB-BASE

The sub-base can be roadbase, crushed rock or concrete. For foot traffic, use 100mm of compacted roadbase. For drives, use 200mm or more of roadbase, depending on your soil type, or 100mm of reinforced concrete. Calculate the area of paving and multiply by the sub-base depth. This will give the volume of material required. For roadbase, multiply volume in m^3 by 2.25 to convert to tonnes.

Spread the sub-base over the area and use a plate compactor to consolidate it. If the depth of sub-base is over 150mm, do this in layers. The height will now be the thickness of the paver plus the depth of the bedding sand below the finished height of the paving.

The best sand for bedding pavers is a gritty angular sand; check with a local supplier. Calculate the area to be paved and multiply by the depth of bedding sand (30–50mm). Multiply by 1.5 to convert to tonnes.

Spread the sand over the sub-base to the required depth. Ensure sand is firm but not totally compacted. The easiest way to achieve a smooth surface is to use screed rails — lengths of square section aluminium or PVC conduit (**1**). Set the rails to the height of the bottom of the pavers, plus 2–3mm to allow for final compacting. Ensure you have allowed a minimum of 1:100 fall on the rails for drainage (**2**). Compact the sand under the screed rails to support the next stage. Place a straight piece of timber on top of the rails and using a side-to-side action screed the sand to the top of the rails (**3**). When the area is completed remove the rails and backfill the grooves (**4**).

LAYING THE PAVERS

Before you start laying the pavers, you will need to set up string lines. First of all work out how much space the pavers will occupy by laying five pavers end to end with a gap of 2mm between each. Measure this distance and use multiples or sub-divisions of this figure to set out the string lines.

Try to use whole units wherever possible to reduce the number of cuts and avoid small pieces. Nothing less than one-quarter of a paver should be used. When laying, drop the pavers into place from just above the sand to avoid disturbing the bed. Remember to leave a gap between each paver. Try to avoid walking over the paving while you are laying it, or it may move. Place boards over the work to distribute your weight if you must walk on it. If laying on a slope, start at the bottom and work your way up.

CUTTING PAVERS

After all the full pavers have been laid it is time to cut in. This is best done with a hired brick saw. Mark the pavers to be cut with a waterproof marker, as the lines will disappear under the blade. Wear protective equipment when using a brick saw — ear muffs, safety glasses and a dust mask are essential. Allow water that comes from the saw to stand overnight so that grit can settle out, and then dispose of it thoughtfully. Do not tip it down the drain.

FINISHING

To stop the job from falling to pieces you will need to lock the sides into place. A collar of concrete along the side of the paving will be sufficient for footpaths. The concrete must be from the sub-base to halfway up the side of paver, don't come up the paver too far as you will see the concrete and this will detract from the overall appearance of the job. For driveways, it is best to lay the edge pavers on a bed of mortar placed directly on the sub-base. To lock the pavers into place, sweep fine sand into the joints. Kiln-dried sands are available, or you can just use clean, dry beach sand. You may need to do this a couple of times before all the joints are full.

Finally, compact the pavers into the bedding sand. On a small job this can be done with a rubber mallet and a block of wood, but on larger jobs hire a plate compactor (checking first to see if a plate compactor should be used with your pavers). Use the compactor over a layer of old carpet to avoid chipping.

DOING THE HARD GRAFT

Wet bed paving

The wet bed method of laying — in which the pavers are fixed in a bed of mortar — is often used with large pavers, which can vary in thickness by around 1–2mm. By laying each paver on a bed of mortar, the variations in thickness will not result in an uneven surface. This is a permanent method of laying, since the pavers cannot easily be removed and replaced. It is the fact that there is little chance of subsidence that makes this method ideal for driveways and swimming pool surrounds. Edge restraints are not necessary.

As with the dry bed method of laying pavers, it is important to have a consistent, non-reactive foundation (page 102). The best option for a sub-base is a reinforced concrete slab 100mm thick (page 106), but you can use 150mm of crushed rock or roadbase instead. If you do lay a slab, it must be flat, with an even depth, and have expansion joints at 6m intervals and where it butts up to buildings or other landscape elements.

BEDDING THE PAVERS

Bed the pavers in a mortar mix prepared from 1 part cement to 4 parts 'brickie's' sand, which contains a small proportion of clay. The clay helps to make the mortar pliable and a little bit sticky so it clings to the trowel during laying. Check with a landscape supplier in your area to find a suitable product. The depth of the mortar bed is not critically important, but 10mm is the minimum. Provided the concrete sub-base has be laid properly (page 106), you will be able to work with a mortar bed of a consistent depth, without having to cope with awkward high or low spots.

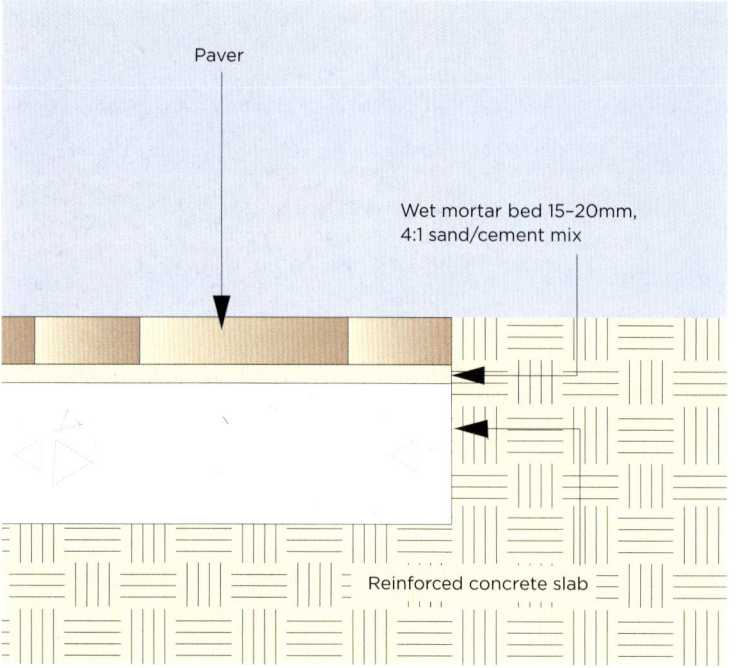

LAYING THE PAVERS

Set up grid lines in the same way as for dry bed paving (page 103), which will act as a reference when laying, helping you to keep everything straight and square. Also position pegs at convenient points around the site to show the finished height of the paving. Start laying the pavers at some convenient point, whenever possible up against the longest straight edge.

1 Put down some mortar. You can either lay a solid bed or position five individual blobs under each paver — at its centre and at each of its four corners. Some people like to use blobs because this method allows the mortar to move into the voids under the paver. When laying with a solid bed, the only place for the mortar to go is out from under the sides.

2 Settle the paver into the mortar with a few gentle taps from a rubber mallet, having checked beforehand that the mallet you are using doesn't leave unsightly marks.

3 Check the paver is at the required levels — flush with those already laid — and that it has the correct fall for drainage. Use your spirit level with a long straightedge so that you can check it against other pavers on all sides. Once all the full pavers have been laid, cut in smaller pieces (page 103).

FINISHING

Depending on the style of paving, you can leave a 2mm-wide gap between pavers, to be filled with fine sand later (page 103), or install 5–10-mm-wide mortar joints. Mortar jointing is the more difficult method, and there are two ways of doing it: grouting, where a paste is worked into the joints with a sponge and the excess wiped off afterwards; and individual jointing, carried out with mortar and a trowel.

Grouting is probably quicker, but not all types of pavers can be grouted — check with the paving supplier or manufacturer — and the job can look messy if it is not done properly. Test your skills on an inconspicuous area before attempting anything ambitious. The most important points to remember are to prepare the grout to a thick creamy consistency, and to sponge the excess mortar off the surfaces of surrounding pavers before it dries. This can be particularly difficult to do where the pavers have a rough finish.

If you are placing mortar into individual joints, you also need to be careful not to get any onto the surface of the pavers, where it will stain. Mix the mortar to a fairly stiff consistency, place some on a board and cut a small piece from the main pile. Roll this into a long cylinder about the thickness of your finger and use a trowel to drop it carefully into the joint, making sure that it fills it completely. Leave the mortar to set slightly, then trim off any excess. You can also use what is known as an 'ironing tool' (available from paving suppliers) to give the joint a neat finish.

DOING THE HARD GRAFT

Concrete paths, drives and paving

Concrete has several advantages over most other types of materials: it can be formed to suit any shape, it is long lasting and can be given a variety of finishes. Among its disadvantages are staining, cracking and surface deterioration over time. Apart from its use as a paving material in its own right, concrete can also be used as a sub-base for various other types of paving (pages 102–105). The preparation of the slab is the same in both cases, with the degree of finishing being the only major difference.

When planning a slab, make sure you allow for expansion or contraction. Provide joints wherever the concrete meets an existing structure, and for paths and drives at regular intervals of around 6m. For finished concrete paths you will also need to place dummy joints every 1.2–1.5m. These are designed to control cracking, and are extremely important if no reinforcing steel has been used in the construction of the path. The concrete is thinner along the grooves, so any movement will cause cracking there, rather than at random elsewhere.

When building a drive, remember that in most cases council approval is needed. Some parts of a drive, such as where it joins the kerb and gutter, known as the layback, are on council land and must conform to regulations. Footpath crossings must also be approved.

PREPARING THE SITE

As is the case with all other types of paving, foundations are very important for concrete. If there is subsidence, the slab will crack and drop. Steel reinforcing will prevent this to some degree, but it is better to start with a solid base. Excavate to a consistent depth equal to the thickness of the slab (say around 100mm), but do not try to use the excavated soil as fill under the slab because it will not be consolidated, and this will lead to subsidence later. If you do need fill, use sand or crushed rock.

FORMWORK AND REINFORCING STEEL

Build the formwork from 100 x 50mm timber, with the top of the form at the finished height of the slab. Take care where sections of formwork join to ensure that surfaces line up, otherwise the concrete will have an unsightly edge. Use plywood that you can bend to form curves. Support the forms with pegs every metre, or even closer together if the formwork is very deep or — as in the case of plywood — not very strong. For deep formwork, consider extra 45° bracing at each peg. Keep in mind that one cubic metre of concrete weighs over two tonnes, so if the formwork moves during pouring you will have little chance of stopping it. Drive pegs below the top of the formwork so you can run a screeding board along its upper surface to form the surface of the slab.

Once the formwork is in place, you can position the reinforcing steel. Use F72 steel mesh reinforcing, allowing 50mm of clear space between the steel and the formwork to prevent rusting later. Support the steel on metal or plastic chairs and use wire to tie the structure firmly together.

POURING THE CONCRETE

In most cases, I recommend that you order premixed concrete. The supplier will deliver to the site and discharge the load either directly into the forms, into barrows or a pump. When ordering, you need to know the number of cubic metres (m^3) of concrete required (the area of the site multiplied by its depth). Concrete is sold in increments of $0.2m^3$. For most domestic jobs, order 20MPa (strength) concrete with an 80mm slump (workability).

Make sure everything is ready the day before you intend to pour and that you have enough help on hand — once the truck arrives you have about 30 minutes to unload before a penalty fee is charged. Place the concrete into the forms and spread it with a shovel, using a chopping action to help move the concrete into all the corners and dispel trapped air. Screed the surface as soon as possible with a sawing action, but do not overwork it. Some water — known as bleed water — will rise to the surface, and this must evaporate before you can finish the slab. Excessive working will only bring extra water to the surface and weaken the finish.

FINISHING OFF

If the slab is to be used as anything but a sub-base it must now be finished. Once the bleed water has evaporated, use an edging tool to give the sides of the slab a curved finish, which will help to reduce chipping when the forms are removed. Scribe dummy joints, if required, and then run a wooden float over the surface to remove air from the slab and any lumps and bumps left after screeding. Finally, if you want to give the surface some grip, gently run a strong bristle broom backwards and forwards across it. Finish off by remaking the edges and the dummy joints.

Many other surface finishes for concrete are possible — such as stencilling, stamping and exposed aggregate — but these call for special skills and experience and are best left to the professionals.

A final word of warning: concreting a driveway is a big job. If you have any doubt about your abilities, call in the professionals.

DOING THE HARD GRAFT

EDGING

Brick, timber or stone edging for a neat finish

Edging is usually used to separate one area in a garden from another — usually lawns from garden beds — but it also has other uses, such as providing a hard surface for the wheels of a mower to run on. Edging also has the secondary function of linking and unifying areas or 'garden rooms', drawing the eye from one 'room' to the next. Once installed, edging should be level with the soil under the turf, and above the level of adjoining garden beds to allow for mulch. Failure to set the edging to the correct height will result in mulch spilling over onto the turf, causing hours of unnecessary maintenance.

INSTALLING TIMBER EDGING

Timber — either H4 treated pine or a selected hardwood — can be used to create an obtrusive edging. Although both types of wood have a limited life span, they can be installed in a fraction of the time required for bricks or pavers.

1 Dig a trench deep enough for the timber, placing the soil on the garden bed side so you can check levels during construction, and set up stringlines for guidance on straight runs.

2 Place the edging timber loosely into place in the trench and hammer in 50 x 50mm hardwood support pegs every 1–1.5m.

3 Drive the pegs below the top of the finished height of the edging, where they can be covered by turf or mulch.

4 Nail or screw the board to the pegs. Place a sledge hammer behind each peg to prevent it from being knocked out of alignment when the nails are being driven in. If a peg must be used on the garden side of the board, cut its top off at a 45° angle to make it less obtrusive. Where boards meet, either make sure a peg bridges the join or cut an extra piece of timber and nail it into place to span the two ends.

5 Backfill the turf side of the timber to 10mm below the top of the board and lower on the garden side to accommodate mulch.

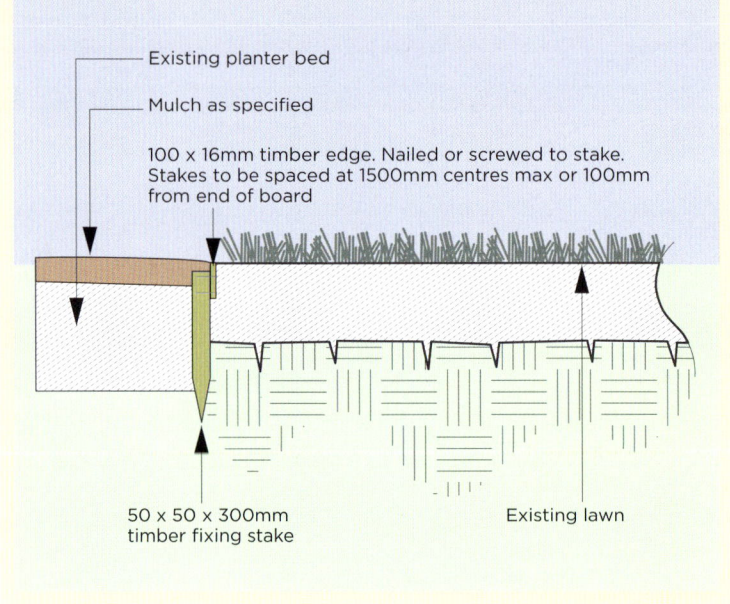

BRICK, STONE OR PAVER EDGING

In general, it is best to have a concrete footing for brick, stone or paver edging, as this will give a solid base. The concrete should be a minimum of 50mm thick and 50mm wider than the bricks or pavers to be used. Work out where you want the top of the edging to come, and then install the footing so that its surface is the depth of the brick or paver (plus 10mm to allow for mortar) below this. Alternatively, you can lay the edging directly onto a bed of thick mortar that has been reinforced with galvanised bricktor (**1**). This is similar to chicken wire, and is sold in rolls 90mm wide. When laid in the mortar, it provides strength and helps to resist soil movement. For straight runs, set out a string line on the garden side of where you want the edging to be. That way, if you lay the edging against it, any variation in the bricks will be hidden by turf and you will see only a clean, straight line on the garden side. Check the level of the edging using a straight edge every metre or so, and remember to stay level with the soil under the turf. Use a piece of PVC conduit to form curves, pegging the conduit at intervals to form the desired line. Lay the bricks or pavers onto a bed of mortar, using a 4:1 sand/cement mix. You can haunch (build a supporting wedge of mortar against the sides of the edging) on the turf side, as this will be concealed, but on the garden side keep the haunch to a minimum otherwise it may become visible when the mulch breaks down. Apply sufficient mortar to each brick or paver to allow a 10mm joint (**2**), which will assist when forming curves. However, if a curve is too tight, the joint will extend on the outside of the bend and look unattractive. In that case, you will need to cut a wedge off the end of each brick to maintain a consistent appearance. Wipe off any mortar spills with a moist sponge (**3**) and 'iron' the joints with a tool which resembles a piece of smooth reinforcing rod. This helps to push mortar into the joints and gives a clean finish.

DOING THE HARD GRAFT

SOFT SURFACES

Alternatives to paving

Too much paving in a garden can sometimes look a bit harsh. Avoid this by using loose materials such as bark chips, pebbles, gravels or crushed stone. They are fairly solid underfoot and serve many of the purposes of solid paving, but are easier to install, softer in appearance and, most importantly when working around existing trees, will allow water and air into the soil beneath.

Gravels and pebbles are best used on level areas or shallow slopes; if the slope is too steep the material will gradually work its way down and require constant maintenance. There are many materials to choose from, and a wide range of local variations. Visit a local supplier to see what is available, keeping in mind how the area is to be used. Larger pebbles are attractive, but difficult to walk on, so use them as features only. Smaller pebbles and gravels are best for traffic areas. In most gardens, loose materials need to be contained by an edging.

When using gravels and pebbles for drives, treat the area in much the same way as for paving. Compact a 100mm bed of crushed rock or roadbase and allow for only 20 or 30mm of gravel on top. If you use too much gravel, it becomes difficult to drive and walk on. When using a light-coloured gravel, place a thin layer of crushed stone — similar in colour to the gravel — over the base first. This will stop darker material beneath from showing through when the gravel is scuffed. Use crushed tile dust with red gravels. For pedestrian areas, you can place the chosen material directly onto the existing soil. However, to prevent the gravel disappearing into the soil over time, first put down a layer of geotextile fabric or weed mat. This will act as a barrier, but will still allow water and air to enter the soil freely.

'Toppings' and decomposed granite can be used when something more solid than pebbles or gravel is called for. In most cases, these materials are made up from small gravelly pieces embedded in finer particles of the same base. Each district has its own version, so check with your local supplier to see what is available. Installation is the same as for gravel and pebbles, but with the addition of some off-white cement in a ratio of 1:10 for stability. Place more material between the edge restraints as it will compact down. On large areas, use a vibrating plate compactor, while on smaller areas simply walking over the surface a few times will do. Lightly hosing with water will help the cement set.

To create a different look, throw fine pebbles or gravel over the surface before compacting. Each time it rains the small gravel will come to the surface and make the area look new again. It is important to have a well-designed drainage system when using toppings. Once water begins to move over the surface it forms runnels and this will eventually lead to unsightly scouring. To maintain the surface, rake over it to cover scuff marks.

For high traffic areas, stepping stones combine the benefits of paving and the look of gravel or turf. Use split stone for a natural feel or large pavers for a more formal look. Begin by roughly placing the stones at regular intervals, about 600mm apart from centre to centre, and then adjust the spacing to suit your natural stride pattern. If you want a natural feel, avoid straight lines by offsetting the stones while keeping the spacing even. For a formal setting, place a stringline along one side of the stones. Excavate to allow for the depth of the stone and a 30–50mm mortar bed. Place five pads of mortar under each stone (pages 104–105) and check that the pavers are all in the same plane. Start by laying the first and last pavers, and then use a straight edge across the top of them to adjust the level of intermediate stones.

Four types of loose surface, showing some of the finishes that are available (clockwise from top left): **black pebbles with timber pontoons; a bed of loose bark chips; white gravel with slate steppers; sandstone pebbles.**

DOING THE HARD GRAFT

IMPROVING YOUR SOIL

Attending to the basics

The type of soil you have, and its suitability for the plants you are trying to grow, will play a large part in determining whether your garden will be a success. There's not much point spending good money on new plants, only to put them in to a sub-standard soil where they will struggle. Of course, you may be one of the lucky ones with a garden full of excellent soil, but if not, it's worth putting in time and effort to improve soil quality.

The aim of improvement work is to get the content, texture and structure of the soil to a point where it will provide plenty of nutrients, water and oxygen to the plants, as well as good, reliable drainage. Testing the pH level of your soil to better assess its suitability for planting is also worthwhile. A pH test kit can be purchased from most nurseries and hardware stores. You hear the term 'a moist, fertile, well-drained soil' used all the time, simply because that's what plants like. Some Australian natives may be an exception to this rule, but for most other purposes you want the best soil you can get.

SANDY SOIL Water drains quickly from sandy soil, taking nutrients with it. To combat this you will need to add not only plenty of organic matter, but also something to slow down the nutrient loss. The best way to do this is to add soil with some clay content, which will help hold water and nutrients for longer. Dig in the additional material as thoroughly and as deeply as you can, ideally to a depth of around 300mm. Sandy soils also tend to repel water after dry periods, which can be remedied by the measures described above, and also by the application of a wetting agent, which is available from hardware stores and nurseries.

CLAY SOIL At the opposite extreme there are clay soils, which hold water and nutrients well, but often don't drain well, especially if compacted. To remedy these defects, dig in organic matter and gypsum to help open up the soil, improving drainage and also the availability of oxygen.

VERY IMPOVERISHED SOIL There are situations — particularly on new housing developments — where most, if not all, of the original topsoil has been stripped away, leaving hard clay soil that is useless for growing. The best way to deal with this situation is to build raised garden beds or planter boxes on top of the existing surface. However, take care to integrate the new soil into what is there by adding it slowly, digging thoroughly as you do so to mix the two layers together. Once this is done, you can start to build the level up to the desired height. If you don't integrate the layers, water won't drain properly and plants will have trouble growing.

MULCHING is important for the success of any garden, and a 50–75mm-thick layer of organic mulch will work wonders. Not only will it reduce water loss through evaporation; but it will also help water to get into the soil; combat soil erosion; feed the soil as it breaks down; act as an insulation barrier to keep plant roots cooler in summer and warmer in winter; and reduce weed growth. And — if that's not enough — it even looks good! There is just one cautionary note: don't build mulch up high around the bottom of your plants, since this restricts air flow and can cause rotting. Thin the mulch down just around plant bases.

DOING THE HARD GRAFT

TURFING

The grass can always be greener

Many factors will influence your choice of grass. In warm areas, where there is little chance of frost, grasses such as buffalo, Durban and couch are best. In colder areas, tall fescue and perennial rye grow well. Some grasses can be established by seed, others only by turf. Turf is more expensive in the short term, but the effect is immediate. Seed is cheaper, but will take longer to establish and can look patchy. Most turf grasses require full sun, so if you have a shaded area it may be better to choose another type of surface. Some types of turf are also better able to handle wear than others. Draw up a list of all the relevant factors and then ask a local supplier for their recommendations. Turf is sold by the square metre (m^2), and comes in 1m^2 rolls. Work out the area to be covered, and add 5 per cent extra for regularly shaped areas and 10 per cent for irregular areas.

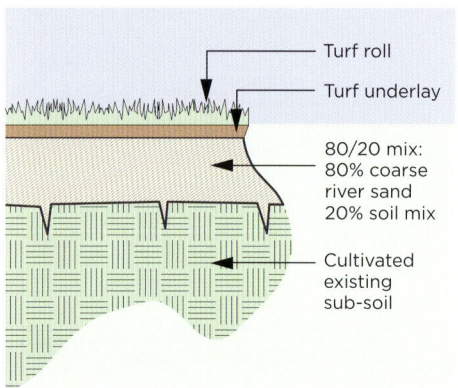

PREPARATION Having chosen the grass, prepare the site. Carry out any soil improvements (pages 112–113), but don't add too much organic matter. Next, consolidate the soil to remove some of its air. On a small area you can do this by walking over it, but hire a water-filled roller for larger areas. The site also needs to be levelled. Use a metal rake or topsoil spreader (above) to remove bumps, spreading the soil and re-consolidating it as you go. You should be able to see a light footprint on the surface when you've finished. Finally, rake in high-phosphate fertiliser and water in.

LAYING Start laying at the bottom of any slope and lay across it, staggering the joints as you go so they don't line up. Keep all the edges butted tight together since gaps will allow the turf to dry out and increase shrinkage. Gaps also lead to an uneven finish on your lawn. Avoid walking on newly laid turf, but if it is unavoidable, lay boards to distribute your weight and minimise the disturbance. If you have calculated accurately, you should have enough turf to avoid using small pieces which dry out quickly and will take longer to establish.

ESTABLISHMENT This is the most critical period for your new lawn. Remember that turf has very short roots so it is unable to collect water from the soil. The area must therefore be kept moist (not wet) until the turf has developed a root system. Check to see if it has established by gently trying to lift a section — you won't be able to if the roots have grown. Do not attempt to mow until roots are established, or you will run the risk of causing irreparable damage. Once the roots have taken, the leaf will start to grow and you can then remove one-third of the new growth at a time.

MAINTENANCE If you want your new lawn to remain looking good you will need to maintain it. Water when necessary, and remember that deep infrequent watering is better than light daily watering. Mow frequently, but only remove one-third of the leaf at a time. Longer grass is better able to cope with stress. Because mowing removes nutrients, you will need to replace them with fertilisers. When the grass is actively growing, apply a general lawn food. If you keep the turf actively growing, and have selected an appropriate grass variety, weeds shouldn't be a problem.

Nige says...
ANTICIPATION

It was a mate of mine named Ross who pointed out to me how smart ants are. He worked as a green keeper on a golf course, and fertilising is a major part of maintaining the greens. Fertiliser always needs to be watered in immediately after it's been applied, and the smart thing to do is to spread it just before rain because this will save time and, more importantly, water. It amazed Ross that his boss always seemed to get it right. If he asked Ross to apply some fertiliser, then sure enough rain would follow a short time later. But how did he know? Ross asked him one day, and he revealed his secret: the ants told him. He reckoned that if you kept an eye on the trees around the course, whenever you saw the ants all coming down and returning to their nests then this was a sure sign of rain to come. Old gardening hands are full of tips and tricks like that, and I'm always interested to hear them. To me it just goes to show the importance of working with nature rather than against her when you're planning and maintaining a garden.

DOING THE HARD GRAFT

PLANTS FOR THE GARDEN

Make sure you have a plan

The biggest problem I see in gardens is easy to avoid. You never go into the supermarket without a list of the items you need, and the same is true of garden centres. Never buy a plant because it looks good and then decide where to put it. You need a list of the plants you are looking for, having thoroughly researched those that fulfil the requirements of your design.

Once you are at the nursery, here are a few tips to ensure you get the best value for your money. First of all, check that the plants you buy are the ones you actually want. There are many varieties and they can differ a lot. For accuracy, use the Latin names of plants, rather than the common names. Any good gardening book will give both common and Latin names. Next, make sure the plants aren't root-bound. The easiest way to check is to tip the plant out of its pot. If the roots circle the outside of the root-ball, or there are roots escaping the drainage holes, don't buy it. Roots that circle stay that way and can stunt and eventually kill a plant. Roots that do not adequately fill the pot can also pose problems, and this can also be a sign that earlier root-bound stock has been re-potted. And finally, carefully check any plant you are considering for insects, diseases or weeds. If you see any, don't make the purchase, otherwise you will just be introducing a new range of problems into your garden.

SELF-SUPPORTING All plants, if they have been well grown, should be self-supporting. Trees in particular need to develop 'trunk taper' — a thickening from the base of the trunk — at an early stage. Plants shouldn't need staking, except for mature specimens that may need some initial support (see diagram and caption below), and a plant with a properly tapering trunk won't need support. Before purchasing stock, remove any supporting cane from a plant to check that it can stand by itself. If it bends dramatically, don't buy it.

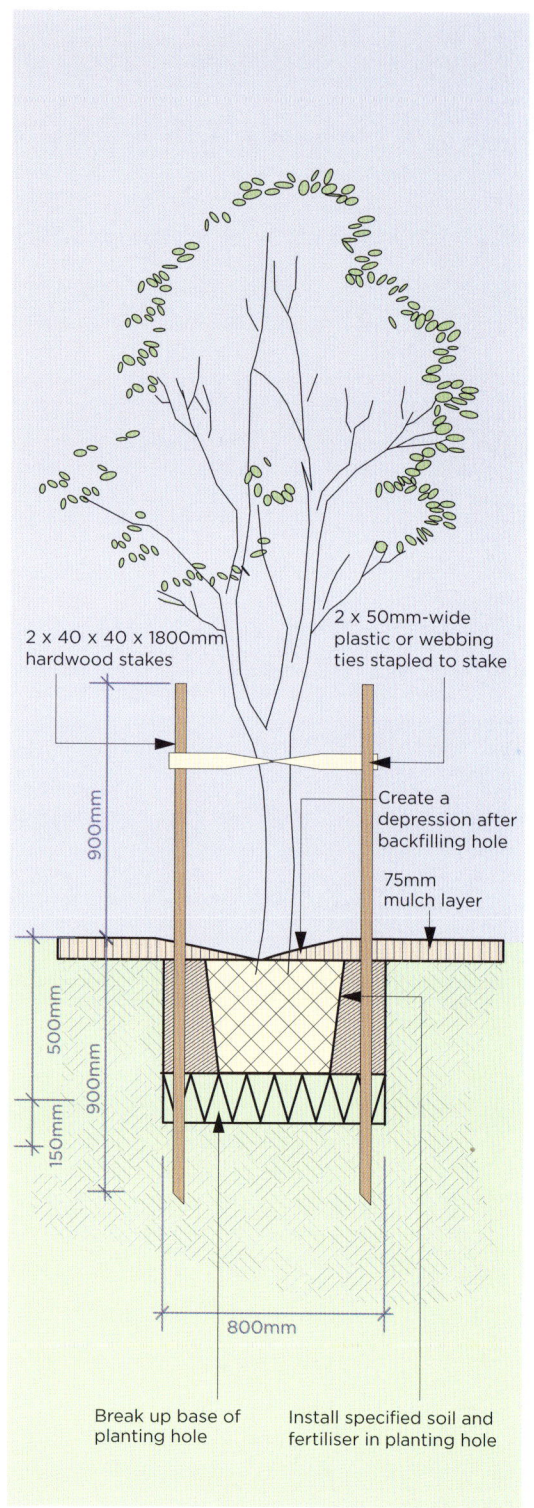

Most plants will not require staking when you put them into your garden. However, this may not be the case with more mature specimens, which can take some time to become self-supporting, especially in windy areas.

PREPARING THE HOLE Having got your purchases home you can prepare for planting. Dig a bowl-shaped hole about twice the volume of the pot to create an area where roots are able to spread out. In heavy clay soils, break up the base of the planting hole to assist with drainage. Avoid digging into subsoil, as this can create a well that will fill with water. Don't discard the soil from the hole. Water stock the day before planting to minimise stress.

PLANTING Remove the plant from its pot, check for abnormalities and if necessary rough up the surface of the root-ball to remove any circling roots. Place the plant into the hole so that the top of root-ball is at the same height as the surrounding soil. Adjust the depth of the hole accordingly. Ensure that the plant is upright, with its best face showing. Some growers mark north, and the plant should be located accordingly to prevent sunburn.

BACKFILLING Use the soil saved from the planting hole as backfill. You may need to add some organic matter, but with large specimens don't dig organic matter deeper than 200–300mm into the soil. Half fill around the root ball and water to allow air pockets to disperse. Fill the rest of the hole and create a well around the root-ball so that water is directed through the root-ball and not just around it. Don't squash the soil into the planting hole.

WATERING IN Water thoroughly, even though the soil may be moist. The plant has not yet developed roots outside the root-ball and water must be directed into the root area to encourage growth. Remember that over-watering can be as damaging as under-watering. Finally, lay a 75mm-thick layer of mulch over the planting area. This suppresses weeds, insulates the soil from temperature extremes, reduces evaporation and adds organic matter to the soil.

DOING THE HARD GRAFT

LOOKING AFTER YOUR NEW GARDEN

Death, taxes and weeding

Your dream garden is now complete, so all that's left to do is sit back and enjoy all the hard work, blood, sweat and tears. Well, not quite — like everything in life, there's maintenance. But don't be alarmed, by observing a few basics the time spent maintaining your creation shouldn't outweigh the enjoyment that you gain from it.

WATERING This is important because plants under stress won't thrive, and this will make them more susceptible to pests and diseases. By keeping soil moist at all times — but never wet — you will reduce the stress on plants. Also mulch to reduce evaporation from the soil and help control temperature fluctuations. Irrigation can cut down the time spent hand watering, and you will find a vast range of irrigation components on sale at garden centres and most large hardware stores. Take care, however, because in many areas irrigation systems must be installed only by a licensed operator, so it is important to check with your local authority first.

WEED GROWTH Changes made to a site during a landscaping project will inevitably result in a flush of weed growth as seeds that have remained dormant for years suddenly spring into life. A thick layer of mulch should keep this outbreak to a manageable level, but ensure you top up the mulch as it decays into the surrounding soil. Weeds need to be controlled before they set seed, and a daily stroll around your garden is the perfect time to pull out the few specimens that have had the audacity to appear overnight. Do this, and you won't have to spend an entire weekend slaving in your garden pulling weeds. For those persistent weeds — such as onion weed, oxalis and nut weed — which seem to survive whatever you do, keep a spray bottle of herbicide handy. Ask your local nursery to recommend a suitable product.

FERTILISER Like people, plants need food, which they obtain from the nutrients found naturally in most soils. However, many Australian soils are deficient in particular nutrients, so you may need to supplement the naturally available food with a fertiliser. There is a huge range of fertilisers on the market, but most fall into one of three groups — chemical, organic and a mix of both. Chemical fertilisers are fast acting, they can address a specific deficiency and are easily applied. However, too much can kill a plant. Organic fertilisers are slower acting since they require organisms in the soil to break them down; they contain a wider range of nutrients, but they can be bulky and smelly. However, they have one major advantage in that it is difficult to over-fertilise with organics. Many complete fertilisers are combinations of chemical and organic, and as a result they have many of the advantages of both. The best time to apply fertiliser is at the end of the first flush of growth, and then again later in the season. All fertilisers have an NPK (nitrogen, phosphorus, potassium) ratio, and often other elements as well. Different plants have different requirements — ask your local nursery for a recommendation.

PESTS AND DISEASES These will normally only affect plants that are under stress, and this can be caused by anything from too much water to poor soil. Therefore, choosing plants that suit your local conditions and soils, preparing the soil and buying healthy stock will all help to limit the occurrence of pests and diseases. If you do happen to notice one or two grubs on a plant, don't be too alarmed since it is generally fairly easy to pick them off by hand. However, if you feel that an infestation is about to overwhelm your garden, take a specimen of whatever it is that is causing you concern to your local nursery and ask for their advice on how to deal with it, preferably without having to resort to a pesticide.

CAN'T DO WITHOUT IT

IF YOU'VE READ THE PRECEDING CHAPTERS CAREFULLY, YOU NOW KNOW THE BASICS OF PLANNING AND CREATING A GARDEN. NOW ALL YOU NEED TO KNOW ABOUT ARE A COUPLE OF THINGS YOU CAN'T DO WITHOUT — THE RIGHT TOOLS, SAFE WORKING PRACTICES AND SOME SIMPLE ARITHMETIC.

CAN'T DO WITHOUT IT

TOOLS OF THE TRADE

There's no substitute for quality

Always buy the best tools you can afford. Good-quality tools not only last longer, but they generally allow you to do a better job as well. You may be able to pick up a cheap spirit level in the bargain bin at your local hardware store, but is it really a bargain if you have to redo a job because the initial measurements were out of square? If you pay good money for tools, it makes sense to look after them. Make sure you clean off dirt, resharpen blunt edges, protect against corrosion and store them away safely after use.

Measuring and marking

Accurate measuring and marking is an essential part of most construction jobs in the garden. Wavy paving surfaces and posts that lean slightly are marks of the amateur. Because all tools receive a few knocks while out of doors, it is vital that you check levels and squares regularly to ensure that they are still accurate.

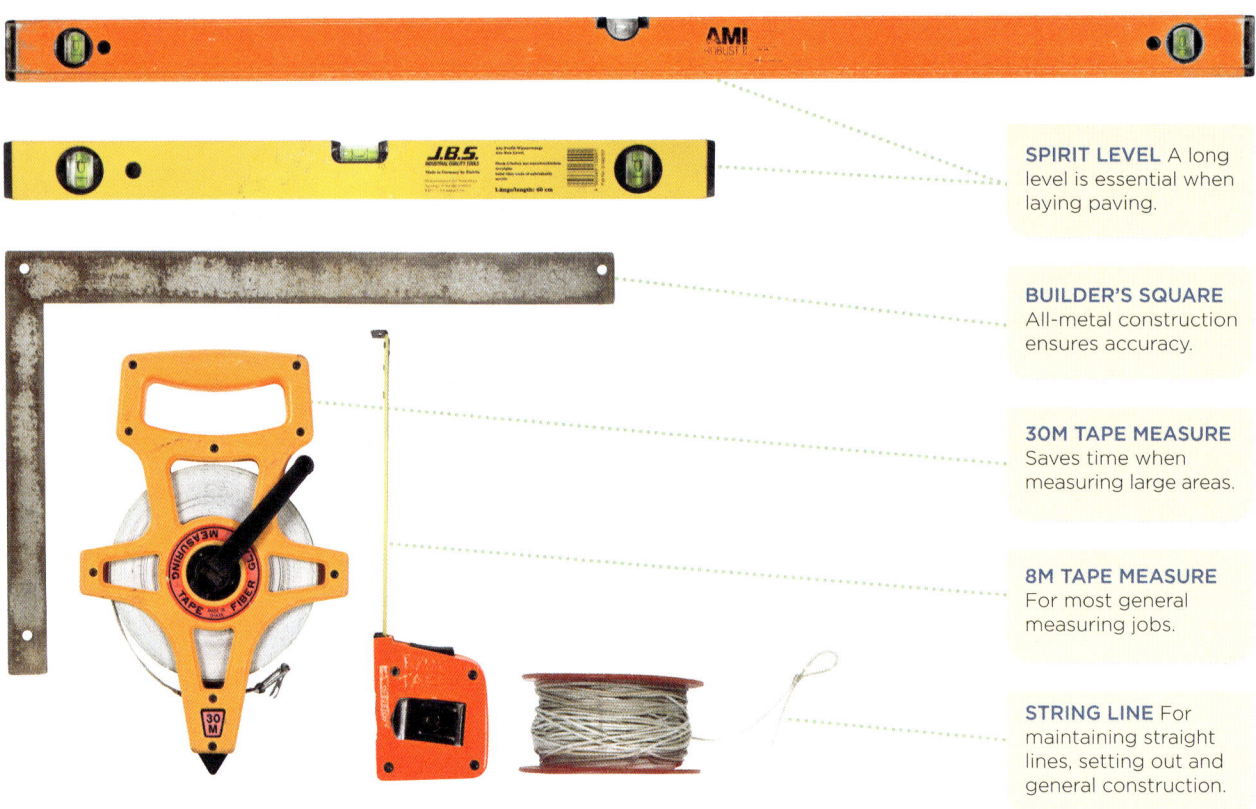

SPIRIT LEVEL A long level is essential when laying paving.

BUILDER'S SQUARE All-metal construction ensures accuracy.

30M TAPE MEASURE Saves time when measuring large areas.

8M TAPE MEASURE For most general measuring jobs.

STRING LINE For maintaining straight lines, setting out and general construction.

CARPENTER'S SQUARE For marking 90° angles.

WOODEN MALLET Soft head prevents damage to chisels.

FOLDING RULER Handy pocket-size rule folds out.

Carpentry

Many home landscaping jobs involve some carpentry, such as building decks and fences for example. The tools you need will vary according to the jobs you have to do, so it is probably best to buy a basic tool kit to start off with, and then add to it as and when you need to. The important point with all cutting tools, such as saws and chisels, is to keep them sharp.

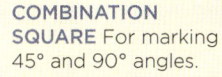
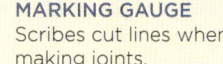

COMBINATION SQUARE For marking 45° and 90° angles.

MARKING GAUGE Scribes cut lines when making joints.

G-CLAMP Holds items temporarily while working or gluing.

WOOD CHISELS For cutting joints and housings.

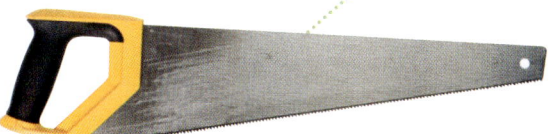

PANEL SAW Used for most general-purpose sawing jobs.

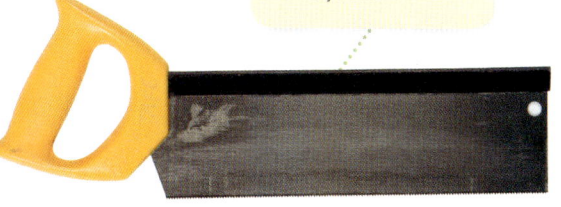

TENON SAW Mainly used for cutting wood joints.

BENCH PLANE For reducing the thickness of wood and finishing.

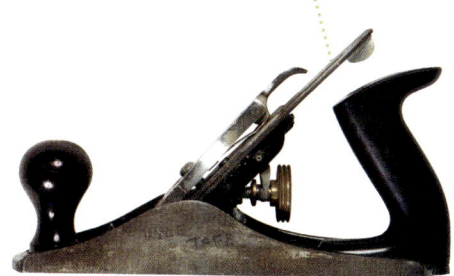

CAN'T DO WITHOUT IT

Brick, stone and paving

Apart from a small range of common tools used for most jobs involving bricks, blocks and pavers, there are also a number of tools designed specifically for working with particular materials. These will make the job easier — once you know how to use them properly — but unless you plan to undertake some major construction work you can probably make do with only the basics to save expense.

Brick

SCUTCH HAMMER Replaceable 'comb' used for shaping.

TROWEL For placing, spreading and shaping mortar.

JOINT RAKE Used to create neat, uniform mortar joints.

BRICK GAUGE Guide for measuring the height of brick courses.

Common tools

LUMP HAMMER Heavy head provides force for cutting.

BOLSTER Wide chisel used for cutting brick, stone and concrete blocks and pavers.

STRING LINE Essential for keeping lines straight when paving or building.

IRONING TOOL Used to shape mortar joints neatly before they set.

Stone

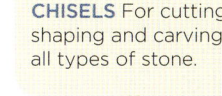

SCUTCH AND POINT HAMMER For rough shaping of stone.

RUBBING BLOCK For smoothing rough surfaces on stone.

WOODEN MALLET Used with chisels for shaping and carving.

WEDGES Used for splitting large blocks into smaller pieces.

CHISELS For cutting, shaping and carving all types of stone.

Paving

RUBBER MALLET Used for seating pavers when they are being laid. Choose a head material that will not leave marks.

STRAIGHTEDGE A long straightedge — one at least 2m long — is essential when laying pavers (not illustrated).

125

CAN'T DO WITHOUT IT

Laying concrete

The range of tools you need for a concreting job will obviously depend on how big the job is, and how you plan to tackle it. Illustrated here are only the basic finishing tools, but obviously you will also need tools for preparing and installing formwork, if there is any (see basic carpentry tools, page 123) and reinforcing steel (hacksaw, bolt cutters, angle grinder, pliers), as well as tools for mixing and placing the concrete. If you are ordering ready-mix, you may need a wheelbarrow for placing the concrete if the truck can't get close enough, as well as a spade to distribute it evenly and drive out any air bubbles. For finishing large areas, you will also need a long screed board and a bull float.

JOINTING TOOL For making joints in large areas of concrete.

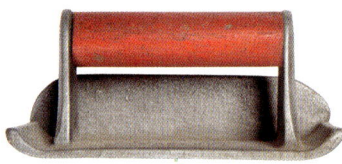

INTERNAL EDGER For finishing internal concrete edges.

EXTERNAL EDGER For finishing external concrete edges.

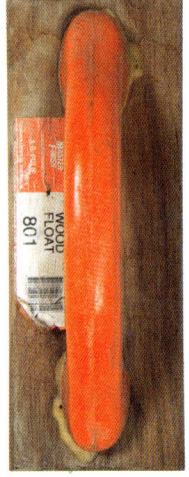

WOOD FLOAT Used to give concrete a rough finish.

SPONGE FLOAT Gives concrete a lightly textured finish.

STEEL FLOAT Gives concrete a very smooth finish.

Power tools

With the huge range of cheap power tools available, there's no excuse for not owning all of the basic equipment shown here (although good-quality tools are preferable if you can afford them). It is generally best to buy mains-powered tools if possible, since many of the cheaper battery models lack power. When working outdoors, make sure mains-powered tools are always plugged into a safety switch.

POWER DRILL Buy mains and portable drills if possible.

ROUTER Plunge routers are a more versatile tool.

CIRCULAR SAW Buy a saw with at least a 190mm diameter blade.

JIGSAW Buy a tool powerful enough for heavy outdoor work.

ANGLE GRINDER For grinding and cutting metal, stone and brick.

SANDER A random orbit sander is the best general tool.

CAN'T DO WITHOUT IT

Gardening

A keen gardener can soon end up with a shed full of tools. Shown here is just the basic equipment that you will need for most home landscaping projects. There is a huge range of special tools available for particular jobs — such as clamshell diggers for post holes — that can make life easier. Check with your local hardware store.

WHEELBARROW (not shown) A vital tool for any garden job. Buy a heavy duty model with timber handles and a pneumatic tyre.

PLASTIC RAKE Used for soft materials, such as leaves.

SPADE The long handle gives more leverage when digging.

STEEL RAKE Used for hard materials, such as sand and gravel.

SHOVEL For moving loose materials, such as soil or gravel.

PINCH BAR Used for levering out roots, rocks and obstructions.

MATTOCK For breaking up hard soil and digging out roots.

FORK Used for turning over loose material such as soil.

Tools for hire

There's no point buying a $2000 machine if it only gets used once every year or so. Hiring is the perfect answer if you find yourself in that situation. If you are unfamiliar with the machine you are hiring make sure you ask for instructions when you pick it up. With large equipment, such as a bobcat, hire an operator as well.

LASER LEVEL WITH MEASURING STAFF AND RECEIVER For establishing accurate levels, particularly when working on large or complex projects. Can be operated alone.

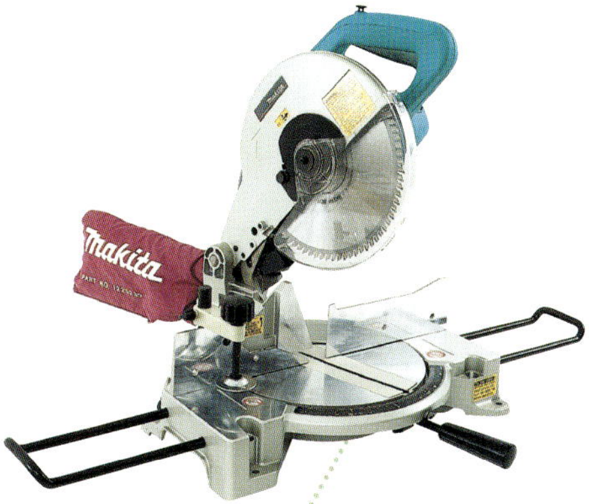

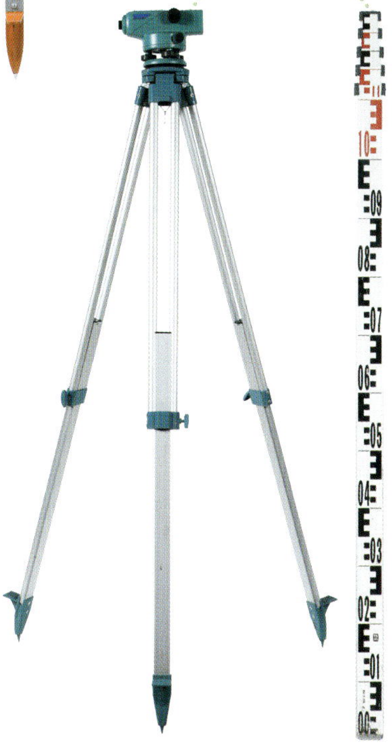

AUTO LEVEL WITH MEASURING STAFF Also for establishing accurate levels on a site. Two people are needed to operate it.

DROP SAW For accurate docking of timber at 45° and 90°.

PLATE COMPACTOR For compacting base material on paths or under paving.

CAN'T DO WITHOUT IT

SAFETY FIRST

Identifying the risks

Safety is obviously a vitally important aspect of any job you do in the garden, or anywhere else around your home for that matter. Being safe on the job will reduce the risk of injuring yourself and others, and help prevent damage to your property and its surroundings. Most important is good old-fashioned common sense. Don't rush in and do something before you've given it careful thought: do I have the skills for this? is that too heavy for me to lift by myself? is the ladder securely anchored? A few simple, basic precautions, backed by a sound knowledge of the tools and equipment you may have to use — particularly power tools and anything you hire that you're not familiar with — and the proper safety equipment, should see you through any project accident free.

Before starting work, take a few minutes to identify and make note of any hazards that have the potential to cause an accident, even if it's your own property and you're very familiar with it. These might include tripping points, slippery surfaces, objects that jut out at head or groin height, uneven ground that could twist an ankle and especially overhead power lines.

In addition to the obvious visible perils, it is particularly important that you locate underground services such as water pipes, power cables, stormwater and sewer pipes, gas pipes, telephone and cable television lines before you start work. Contact the Dial Before You Dig service (phone: 1100, internet: www.dialbeforeyoudig.com.au), and the relevant service suppliers for the necessary information. Even when you've got the information, still take a few minutes to poke around by yourself for anything that may not be on their plans. Home owners often make minor changes to services themselves — such as re-routing a pipe or wire — and a few years later will have completely forgotten what was done.

Once you know where the services are, clearly identify their position with spray marking paint or safety tape, or expose them by carefully removing soil. It's also a good idea to have the names and phone numbers of a few local tradespeople at hand just in case — it saves vital time and a panic stricken search through the phone book if there is an emergency.

Nige says...

SHOCK THERAPY I've put a shovel or mattock through every type of service there is, but the most dangerous and expensive incident happened a few years ago when I was working for a mate of mine. I was digging a large hole for a plant, it was raining, and the hole had a little water it in, which meant I couldn't see the bottom. About a metre below the surface I hit something hard with my shovel and my first instinct was to feel the obstruction with my hand — a natural impulse that very nearly cost me my life. I'd cut through an electrical conduit and exposed the wires inside, and although my fingers only brushed them, that was enough to deliver a powerful electric shock. Luckily the shock didn't cause any lasting damage, but the thought of what might have happened has remained with me ever since and has made me extra cautious, especially where electricity is concerned. The damage cost about $700 to repair, and luckily the client was a good bloke and went halves, so it was an expensive lesson in the importance of thinking safely at all times.

Clothing and equipment

What you wear is up to you, but it's worth remembering that appropriate clothing can reduce the chances of accident and injury, particularly if you don't work outside on a regular basis. Even basic items such as steel- or hard-capped boots — especially if you are handling hard or heavy materials — and clothing with good ultraviolet resistance, a hat, sunscreen and gloves can provide at least some basic protection.

Many items of protective clothing and equipment are job specific, and should always be worn when using particular tools and machines. Essential safety equipment includes: ear muffs, safety glasses, a dust mask (or a respirator with a filter cartridge suitable for dealing with the type of vapour or dust particles involved), a hard hat, knee pads, chainsaw safety pants, gloves and a chemical-resistant safety suit.

Back injuries are common in any kind of physically demanding occupation, and landscaping can certainly be that at times! However, the risk of injuries can be drastically reduced by taking a few basic precautions. When lifting, always keep your back straight, bend your knees, have the object close to your body and use your legs for power. If you want to see how to do it properly, watch a professional weightlifter in action. Never attempt to lift something that is remotely close to your physical limit. If it's too heavy, share the lifting and carrying with another person, even if that means postponing the job until someone is available. You may be impatient to get the job done, but it only takes a couple of careless seconds to do damage that may take many painful weeks or even months to repair. Stretching or twisting while lifting also puts major stress on your back, even when what appear to be relatively light objects are involved. Try to avoid it.

Similar rules apply when shovelling bulk materials, such as soil or sand. Bend your knees and keep your

back straight. A long-handled shovel is a good tool to use for this job. With your knees bent, load the mouth of the shovel, support the handle on your thigh to take the weight of the material, and use this as a pivoting point while your arms control the direction and placement of the load. If you are shovelling material into a wheelbarrow, make sure the barrow is facing in the direction of travel when full. It is much harder to control a fully laden wheelbarrow when you are reversing and turning it, than it is an empty one. It is in situations like this where a little forethought can prevent a lot of wasted effort and the ever-increasing possibility of an injury as the day progresses and your body becomes more fatigued.

Of course, getting someone else to do all the heavy work for you — preferably someone young, fit and built like a tank — will eliminate all possibilities of injury, soreness and fatigue. I used to have a couple of great guys working for me who perfectly fitted that bill. Ant and Jonno could shovel and barrow all day, any day, in any conditions. Best of all they loved it, and so did I. Mind you it used to cost me a fair whack on a Friday night. Their legendary ability to work was more than matched by their legendary ability to drink!

CAN'T DO WITHOUT IT

FACTS, FIGURES AND REFERENCE

Formulas and calculations

Landscaping demands some skill with numbers, but not much beyond the simple arithmetic and geometry you probably learned at primary school. Although most of the calculations are generally fairly straightforward, it is nevertheless important that you take care to be accurate, especially at the planning stage when you are trying to work out the quantities of materials required, and therefore how much the job will cost, and later on at the lay-out stage when you are transferring your plans to the garden itself. Always use a consistent set of units — preferably metric units — and ask someone else to check your calculations at each stage. If there are errors, it is important to catch them well before you start work, not afterwards.

CONVERSION FACTORS

IMPERIAL TO METRIC

Length: multiply INCHES × 25.4 to get MILLIMETRES

Length: multiply FEET × 0.3048 to get METRES

Weight: multiply OUNCES × 28.35 to get GRAMS

Area: multiply SQUARE FEET × 0.093 to get SQUARE METRES

Volume: multiply CUBIC FEET × 0.0283 to get CUBIC METRES

Liquids: multiply PINTS × 0.568 to get LITRES

AREAS OF PLANE FIGURES

Figure		Area
SQUARE	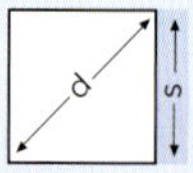	S^2 or $\dfrac{d^2}{2}$
RECTANGLE	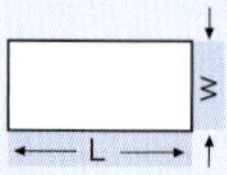	Lw
TRIANGLE	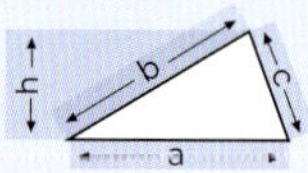	$\dfrac{ah}{2}$ or $\sqrt{s(s-a)(s-b)(s-c)}$ where $s = \frac{1}{2}$ perimeter
CIRCLE	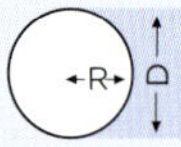	$\dfrac{11}{14} D^2$ or $0.7857 D^2$ or $= R^2$

VOLUMES AND SURFACE AREAS OF SOLIDS

	Figure	Volume	Surface Area
ANY PRISM	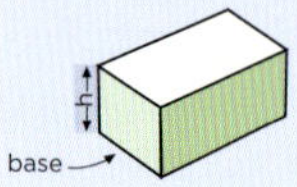	area of base × height (h)	(side faces) perimeter base × height
CUBE	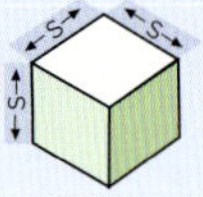	S^3	(whole area) $6S^2$
RECTANGULAR PRISM	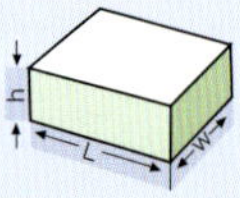	Lwh	(whole area) $2(Lw+Lh+wh)$
CYLINDER	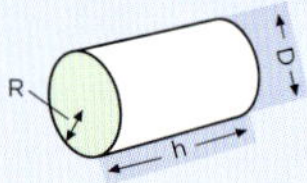	$\dfrac{11}{14} D^2 h$ or $0.7857 D^2 h$ or $\pi R^2 h$	(curved surface) $= Dh$
SPHERE	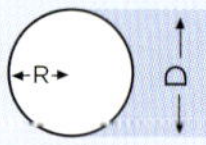	$\dfrac{\pi}{6} D^3$ or $0.5238 D^3$ or $\dfrac{4}{6}\pi R^3$	πD^2 or $4\pi R^2$
ANY PYRAMID	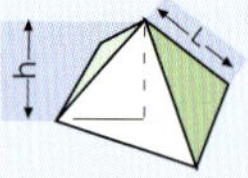	½ Area of base × height (h)	(sloping faces) perimeter of base × ½ slant height (L)

VOLUME AND WEIGHT OF MATERIALS

When working out the quantities of materials needed for a job — sand or gravel, for example — you will end up with a total volume in cubic metres. These figures often need to be converted to tonnes for ordering purposes, and also to give you an idea of how much work will be involved in moving the materials from place to place.

One cubic metre	Weight (in tonnes)
Mulch (organic compostable, approx.)	0.30
Water	1.00
Soil	1.30
Sand (loose)	1.50
Gravel (loose)	1.50
Roadbase (loose)	1.75
Limestone	2.00
Brick	2.10
Sandstone	2.20
Roadbase (compacted)	2.25
Concrete	2.30
Granite	2.60

BUYING LANDSCAPING MATERIALS

You will need to ring around when ordering materials to see what is available in your area. Listed below are some common materials, with details of a few widely used standard products.

- Timber is sold in lineal metres in standard lengths. The shortest commercial length is 900mm, with lengths increasing in 300mm increments (900, 1200, 1500, 1800mm long etc).
- Treated pine sleepers 200 x 75mm are sold in 2.4, 2.7 and 3.0m lengths. Round logs — 100, 125, 150 and 200mm in diameter — are sold in lengths of 1.8, 2.4, 3.0 and 3.6m.
- A standard Australian brick measures 230 x 110 x 76mm. Approximately 250 standard bricks weigh 1 tonne.
- Pavers are sold in many sizes, but one of the most useful is 230 x 115mm because its width is exactly twice its length.
- Pre-mixed concrete is ordered in 0.2m³ increments, with a standard load containing about 5.8m³. One cubic metre of concrete comprises about 30 large wheelbarrow loads.

LANDSCAPING ON THE WEB

The internet is a great source of information on all aspects of landscaping. Listed below is a selection of sites you may find useful.

www.aila.org.au Australian Institute of Landscape Architects

www.outdoordesign.com.au Landscape Industries Association of Australia Inc

www.lcansw.com.au The Landscape Contractors' Association of NSW Ltd

www.qali.asn.au Queensland Association of Landscape Industries Inc

www.lsa.org.au Landscape Association of South Australia Inc

www.liav.com.au Landscape Industries Association of Victoria Inc

www.landscapewa.com.au Landscape Industries Association of WA

www.anbg.gov.au Australian National Botanic Gardens

http://farrer.riv.csu.edu.au/ASGAP Association of Societies for Growing Australian Plants

www.anbg.gov.au/gnp Growing Native Plants

www.opengarden.org.au Australian Open Garden Scheme

www.anbg.gov.au/cpbr/herbarium Australian National Herbarium

www.yates.com.au Yates

www.abc.net.au/gardening Gardening Australia

www.burkesbackyard.com.au Burke's Backyard

www.burkesbackyard.com.au/blitz/home Backyard Blitz

INDEX

a

3:4:5 triangle 75
Acid and alkaline soils 54-5
Agricultural drain 85
Analysing
    a garden site 50-1
    soil 55
Angle of repose, of soil 94
Areas, calculating 132
Asian style
    in tropical garden design 38-9
    influences in a garden 24-7
    Japanese gardens 40-1
Aspect, importance of 52-3
Australian native gardens 36-7
Auto level 77, 129

b

Balcony gardens 60-1
Blob footings 87
Bobcat, hiring 72, 80
Brick
    edging 109
    footings 86
    laying tools 124
    walls 98

c

Carpentry tools 123
Cell drain 82
Chalk line 74
Channel drain, grated 84
Children in the garden 56-7
Circles, marking out 75
Classic garden styles 33-45
Clay soils 54-5, 113
Clearing a site 72-3
Climate, important for gardens 52-3
Clothing, safety 131
Concrete
    footings 87-8
    constructing 88
    types of 87
    mixes 88
    paths, drives and paving 106-7
    composition of 88
    ordering 88
Concrete block walls 98
Concreting tools 126
Contemporary style gardens 42-3
Costs, of garden landscaping 48-9
Council regulations 67
Courtyard gardens 6-11, 60-1
Creating a garden 4-29
Curves, marking out 75
Cutting pavers 103

d

Desire lines 90
Dial Before You Dig
    service 70, 130
Diseases, of garden plants 119
Dish drain 84
Dog toilet 57
Drainage
    behind walls 94
    garden 82-5
    types 82-5
Dry bed paving 102-3

e

Earthworks 80-1
Edging, garden 108-9
Energy efficient garden 18-23
English formal garden 34-5
Equipment, safety 131
Excavations 80-1
Excavators, hiring 80
Expansion joints, for concrete 106

f

Fences 94-9
Fertilisers, choosing and using 119
Footings 86-9
Formal garden style 34-5
Formwork, installing 89, 107
Foundations 86-9

g

Garden
    maintenance 118-19
    planning 30-67
    styles 33-45
Gardening tools 128
Grass, turf 114-15
Gravel surfaces 110

h

Hard graft 69-119
Hiring tools 129

i

Impoverished soils 113
Insurance 67

j

Japanese style gardens 40-1

l

Laser level 77, 129
Lawns
    creating 114-15
    maintaining 115
Legal aspects of landscaping 66-7
Let's make a plan 31-67
Levels, establishing 76-7
Lighting 9

m

Maintaining a garden 118-19
Masonry walls 98
Measuring and marking tools 122
Mediterranean style gardens 44-5
Modular wall systems 96
Mulching 113

n
Native gardens 36–7

o
Order of work, establishing 70–1

p
Paths 100–5
Paver edging 109
Paving
   techniques 100–5
   tools 125
Pebble surfaces 110
Personal style 46–7
Pests in the garden 119
Pets in the garden 56–7
Pit drain 84
Planning
   a garden 30–67
   garden projects 70–1
Plans, garden, creating 62–5
Planting 116–17
Plants, choosing 116–17
Plate compactor 129
Pluses and minuses of a site 50–1
Pollution control 67
Pools and ponds 20–1, 58–9
Post and rail fence, building 98–9
Post and whaling wall 96–7
Power tools 127
Protective eyewear, importance of 43

r
Raft footings 87
Rainfall map 53
Rammed earth walls 14–16
Readymix concrete 88
Red tape 66–7
Retaining walls 94–7
Riser, on steps 90
Rooms, garden 108
Rubble drain 85

s
Safety
   clothing 131
   equipment 131
   identifying hazards 130
   in the garden 130–1
   when working 70, 130–1
   with water 58
Sandy soils 54–5, 113
Scale drawing techniques 62
Setting out a plan 74–9
Site analysis 50–1
Skip bins 72
Small gardens 60–1
Soakaway pit 83
Soft surfaces 110–11
Soil
   disposing of 80
   importance of 54–5
   improvements to 112–13
   quality of 55
   structure of 54–5
Sources of inspiration 33
Spirit level 77
Spray can marker 74
Squares, types of 75
Steel reinforcing, installing 89, 107
Stepping stones 110
Steps
   building 92–3
   garden 90–3
   sleeper, building 93
   stone block, building 92
Stone
   footings 86
   mason's tools 124–5
Stormwater drains 82
Straight lines, marking out 74
Stretcher pattern 100
String line 74
String line level 77
Strip footings 87

Style, developing your own 46–7
Sub-base for paving 102
Sub-surface water drains 82–3
Sunshine hours map 53
Surface water drains 84–5
Surfaces, soft 110–11
Swimming pools 58

t
Television, working on 52
Temperature maps 53
Timber
   edging 109
   fence, building 98–9
   retaining wall, building 96–7
Tools
   for hire 129
   for landscaping 122–9
Toppings 112
Tread, on steps 90
Treated pine, using 93
Trench drain 83
Tropical-style gardens 38–9
Turfing 114–15

v
Vertical, checking for 75
Volumes, calculating 132

w
Walls 94–9
Water
   in the garden 58–9
   level 77
Watering, garden 118
Weather, importance for gardens 52–3
Weeds, dealing with 119
Wet bed paving 104–5
Whalings, retaining wall 96
Wind, effects of 12–7

z
Zen Buddhism, influence on gardens 40

ACKNOWLEDGEMENTS

There are many people who have helped me in one way or another, all equally important and very much appreciated, and this page is dedicated to you. The names are not in any particular order and if I've missed anyone I am truly sorry, but let me know and I'll thank you in person the next time I see you.

My Mum, Dad and brother Steve who I don't get to see that much but have known me longer than anyone else and always been very supportive, even when I was a punk rocker.

Everyone who has helped with this book and believed in me. Jamie for his friendship, encouragement and enthusiasm, and for helping kick-start the whole project, as well as being the publisher — a handy thing to have for a book. Scotty for his friendship, advice and encouragement. Don and Marea Burke for all your help and advice along the way.

Edwin Barnard, well what can I say mate, you've been a massive help in every way from making sense of things to giving me a few more good camping spots. I'll miss the regular meetings and discussions, in particular when the conversation went off on a completely irrelevant tangent and we ended up in hysterics.

John Monty for your teaching ability when I was at Ryde TAFE, and more recently for your knowledge, friendship, expertise and help with the book. To all my teachers at Ryde TAFE.

Everyone at Jamie Durie Publishing and Patio who have worked so hard to put this book together. All of you have your own very special individual talents; one common to all is the ability to tolerate authors such as myself, an achievement certainly worth mentioning. Nicci Hartley, Amanda Emmerson, Belinda Smithyman, Bettina Hodgson, Criena Court, Sonia McAllan, Kirsty Bruce, Nadine Bush, Giselle Barron, Grace Mansour, Julian Brady, Stephen Wells, Dejana Milinkovic, Michelle Kavanagh and Marcus Hay.

The team at HarperCollins*Publishers*, in particular Robert Gorman, Jim Demetriou, Cristina Lee and Robyn Fritchley.

Jason Busch for your excellent photography. I've always believed a picture says a thousand words but some of yours say a hell of a lot more, it's a shame that we couldn't use them all!

I owe a big thank you to all the landscapers who provided gardens and sites to work at; Anthony and Jonno at Entire Landcsapes, Mat, Mike, Rex and Tommy at Land Art, Hamish at Greenfinger Landscapes, Dean Herald and the guys at Rolling Stone Landscapes, Mat Cantwell at Secret Gardens of Sydney, Jim Fogarty at Jim Fogarty Design, Steve Young, Tom and Andy at Balistyle, and of course Nigel Ruck at Garden Life Landscapes.

To all the owners of the gardens used in the shooting of this book. Thanks to Rob Kavo, Shane and Doug (remember boys, don't leave any s**t lying around in the garden!).

All the staff at the North Steyne Emporio for the coffee and great service, and also for giving me an excuse to get out the house and away from the laptop.

Everyone at Channel Nine, past and present who've helped and supported me. *Backyard Blitz** and Lifestyle in general, especially Tim Wise for helping me get into all of this in the first place, and of course Stuart Clark.

My work mates and friends at CTC Productions and *Backyard Blitz** for all their support, friendship, help and laughs over the years; Don, Marea, Jamie, Scotty, Jody, Vid, Niall, Steve, Rick, Andy, Deano, Nicos, Rob Joseph, Rob Beck, Chook, Tom, Lionel, Mitch, Nina, Beth, Jacq, Chris, Mark, Wendy, Jenny, Craicer, Maurice, Adam, Julian, Grant, Brento, Denise, Wendy and all the magazine team. The Land Art boys again.

Hans, Sivina, Garth and anyone else who's helped out behind the scenes.

To Rosie for your encouragement, ideas, those notes you gave me and of course the food! (You too Bec.)

A special thanks to Lily who won't be able to read the book just yet, but will certainly enjoy the pictures — especially the ones she's in. And to Leigh whose love, help and patience are always there.

Finally a huge thanks to all our viewers and supporters, and in particular the smart ones among you who've bought this book. Thank you.

THIS IS THE END OF THE BOOK.